Amateur Radio

General Class Licensing

for 2019 through 2023

License Examinations

Stephen Horan, NM4SH

I0477482

US Amateur Radio Licensing Series

**Amateur Radio General Class Licensing
for 2019 through 2023 License Examinations**

Fourth printing, with updates: April 2022

Paperback ISBN-13: 978-1796686678

Cover image by the author.

Author

Stephen Horan, NM4SH, has been a licensed amateur for over 25 years. During that time, he helped develop and lead weekend licensing classes with the Mesilla Valley Radio Club in Las Cruces, NM. Now he mostly operates on digital modes from his home in Virginia where he is member #1501 of the PODXS 070 Club, has the Loyal Order of Narrow-banded Phase-shifters (LONP) certificate #266, and is an ARRL-certified Volunteer Examiner.

Steve's professional experience includes 23 years as a professor of Electrical Engineering at New Mexico State University and 12 years as an engineer for NASA including four years as a RF spectrum manager. In addition to this series of Amateur Radio licensing study guides, he is the author of the textbook *Introduction to PCM Telemetering Systems* and has published numerous technical articles in journals and conferences.

Visit his author page at https://amazon.com/author/shoran. You can reach him via e-mail at nm4sh@arrl.net.

Also by the Author of this Amateur Radio Licensing Series:

- Amateur Radio Extra Class Licensing

- Amateur Radio Technician Class Licensing

Contents

List of Figures

List of Tables

BACKGROUND

Note to Readers

Welcome back for the next licensing step as you advance your knowledge in the Amateur Radio Service. Your increased knowledge will permit greater access to the amateur bands. The amateur community will also expect you to demonstrate higher technical and operating skills. While the Technician Class license grants some limited High Frequency (HF) access, the General Class license has traditionally been the gateway license for most amateurs venturing out onto the HF bands.

As with the Technician Class license examination, the license examination uses standardized questions, so everyone seeking an Amateur Radio Service license in the United States (US) will see questions drawn from the same master question pool. Amateurs participating in the National Conference of Volunteer Examiner Coordinators (NCVEC) (http://www.ncvec.org) design and maintain the question pool. This version of the question pool is valid from 1 July 2019 through 30 June 2023. The NCVEC will post any changes to the question pool on their Web site. I have designed this study guide to help you through the General Class license examination. The administrators for your license examination will be amateur radio volunteers, usually local, certified to give the test. They have all been through this same process. You will be among friends.

As with the Technician Class examination preparation, this study guide aims to enable currently-licensed amateurs with a basic knowledge of electronics and amateur radio to pass the mid-level General Class examination. Because this is a study guide directed at the examination questions, there continues to be a certain level of compromise in the presentation. I assume that you have basic knowledge and operational experience as a Technician Class amateur, and you desire to improve your knowledge and skills. As you know through your experience as a Technician Class license holder, there is always more to learn about radio science, radio operations, and the Federal Communications Commission (FCC) rules. That will continue after you earn your General Class license.

As with the Technician Class study guide, each chapter covers one part of the General Class question pool. The chapter starts with a short discussion of key concepts behind several of the questions in the chapter that need more context than the space available for the questions' responses. Then, we introduce you to the questions and answers with the correct answer identified, and we explain why it is correct. In many cases, the answer's explanation goes beyond merely picking the correct answer among the options given. To help you learn about amateur radio while we go through the questions, we also include reasons why many incorrect options are wrong. I am striving to provide sufficient instruction through the process to ensure successful completion of the General Class exam element and learn more advanced amateur radio techniques. Once you have successfully passed the examination and have upgraded your privileges, I encourage you to pursue the theory and practice of amateur radio more deeply and work towards the Extra Class license that gives you greater operating privileges.

"How to Study" Suggestions

Given that you are doing this preparation in addition to your other activities, I offer the following suggestions for efficiently using your time.

Become familiar with the material   Before you can study effectively, you need to know where you are going! To do this, look over the question pool and the explanations. The general format is the same on all amateur radio exam elements. There are "technical parts," "operational parts," "safety parts," and "rules and regulations parts." Get a general sense of the flow of the material and the level of detail required. For the General Class exam, the level of electronics and mathematics builds on the Technician Class license knowledge, but you still do not need an engineering degree to pass.

Assess what you already know and what is new   Based on your experience and knowledge of amateur radio as a Technician Class license holder, parts of the material will look familiar. Other aspects, perhaps the complex technical concepts, will be new to you. For now, try to identify those specific concepts that are new to you and place most of your energy there. Also, be sure you understand any specific amateur radio nuances in the familiar material.

Prioritize what is new to you   Depending upon your background and interests, you will grasp some of the new concepts quickly while others will leave you wondering what it's all about. Sort the new topic areas by the level of difficulty to you. Give yourself some confidence by working through the concepts that seem easier to you and building towards the complex concepts.

Make a list of questions Write down the concepts that are difficult for you. Consult resources to which you have access. For example, you can use amateur radio friends, Web references in the study guide, or Wikipedia entries to help you understand the question and the correct answer. If there is an amateur radio club in your area, members will be ready to help you.

Make study aids You can use study aids to help you track the key concepts that you need to learn and help you to become familiar with the material. You can learn items, such as frequency allocations, through experience or memorization. Since you do not have full band operating privileges at this time, you will most likely need to memorize them. To make the memorization easier, make a set of index cards with the allowable frequencies, power limits, and types of emissions allowed. You can also do this with the necessary equations for antenna design, power densities, or other concepts that you are trying to master. Review the index cards a few times each day until you feel more comfortable with the ideas. Do not try to memorize everything in one sitting.

Test yourself You can take one of the online practice exams to see if you are ready for the real thing. Visit sites such as https://www.qrz.com/hamtest/, http://www.arrl.org/exam-practice, or http://www.eham.net/exams/. You can find more sites by using a search engine in your Web browser. When you score consistently above 75 % on the practice tests, you may be ready.

Relax Remember, you do not need a score of 100 % to pass the exam! Do as well as you can in learning the concepts you can grasp. The commentary on the questions will help you understand which of the four choices for the answer is correct and which is incorrect. Learn to recognize the distraction options you can eliminate by using the comments for the question pool as you study. Do not put yourself under pressure to memorize everything. Be willing to tell yourself that you can write off a few questions that you cannot get at this time and hope for the best when you take the actual exam. As you operate with your new privileges, some concepts will become more understandable.

About the Exam

The FCC mandates that the question pool for the amateur service license examinations have a certain structure. The NCVEC has a question pool committee that designs and publishes the questions used in the exam. The general method is to break the overall question pool for all license levels into three major elements covering radio theory, operations, safety, and regulations. The three major license elements are
Element 2 — Technician Class License
Element 3 — General Class License

Element 4 — Extra Class License
Note: there no longer is an Element 1; that was the Morse code examination.

The question pool designers then break the license elements into ten subelements. The designers divide the subelements into question groups reflecting the question pool designers' curriculum. The General Class question pool has over 450 questions in it. Each subelement has between two and five groups of at least ten questions. Each license exam for Elements 2, 3, and 4 will use one and only one question from each group regardless of the number of questions in the question pool for that group. Use this knowledge to help you to design your studies. It would be best if you did not try to memorize the answer to each question but learned the general principles.

Table 1 shows how the designers organized the question pool for the Element 3 General Class license examination. The General exam has 35 questions, and the test designers randomly draw one question from each group. In the individual chapters of this study guide, where we look at the questions, you will see that the start of each question has a code that looks like **G1A01**. We decode this marking as

G1 — Element 3 (General), Subelement 1
A — Group A of Subelement 1
01 — Question 1 from Group A

This study guide uses the same format as the published question pool. On the actual exam that you take, the wording of the questions and the answer choices will be the same, but the answer choice order will be randomized. The order of the question groups within the element will be randomized also. Before selecting the right answer, read each question and the four possible answers carefully.

Before the Exam Day

A Volunteer Examiner (VE) team administers license exams at prescheduled, publicly announced test sessions. Often, you will need to preregister with the team before taking the exam. Be sure to understand the date, time, location, and registration requirements, including any fees, for the session. Ask if this will be an "in-person" or "video-supervised" exam. If it is the latter, ask what equipment you need to take the exam.

If you require accommodations for any special needs, be sure to self-identify to the VE team the type of need and the necessary accommodations you require when you register for the exam. The VE team should permit you to have an accompanying adult or support animal at the session if needed. You may need to work the availability of sound-blocking devices or special room requirements with the VE team in advance.

If you do not have a FCC Registration Number (FRN), you must get one online from the FCC at https://apps.fcc.gov/cores/userLogin.do before the exam. *Note:* you will need to have an e-mail address for this.

Table 1: 2019 – 2023 General Question Pool Organization

Subelement	Content	Groups	Questions
G1	Commission's rules: control operator frequency privileges, good engineering and amateur practice, transmission regulations, VE and VEC, control categories	5	64
G2	Operating procedures: phone operations, operating courtesy, CW operating, volunteer monitoring program, digital operations	5	60
G3	Radio wave propagation: solar effects, atmospheric propagation, ionospheric structure	3	36
G4	Amateur radio practices: station operations, test equipment, interference and grounding, speech processing, mobile operations	5	67
G5	Electrical principles: component characteristics, power circuits, series and parallel circuits	3	43
G6	Circuit components: discrete analog components, digital components	2	27
G7	Practical circuits: power supplies, digital circuits, amplifiers and oscillators, receivers and transmitters	3	40
G8	Signals and emissions: carriers and modulation, frequency domain concepts, analog and digital emissions	3	38
G9	Antennas and feed lines: antenna feed lines, antnna types, specialized antennas	4	54
G0	Electrical & RF safety: RF safety principles, station safety	2	25

Exam Day

The exam will only cover the Element 3 questions from the General Class license pool. You can also take the Element 4 Extra Class exam after passing the General Class license exam if you wish to try it.

You will need to complete the NCVEC Quick-Form 605 at the session; see `http://www.arrl.org/files/file/VEs/605%20Form_2020_Fully%20Inte ractive.pdf` and `http://www.arrl.org/fcc-qualification-question`. To complete this form, you will need to bring
- a copy of your current license
- your valid e-mail address
- your FRN

What else do you need to bring for the exam? Be sure to have the following items physically with you when the exam starts:
- several sharpened pencils and an eraser
- the VE examination fee to cover the expenses of the exam; Note: the VEs receive no pay from this fee or any other fees
- a photo-ID or other valid forms of identification (ask the examination team what alternatives they accept if a photo-ID, such as a driver's license, is not available)
- you may also wish to bring a calculator whose memory you can prove you have cleared; be sure that there are no exam-related formulas or data stored in the calculator's memory or programming because the examination team will check to see that it is clear
- you generally will not be allowed to use cell phones, smartphones, tablet computers, or similar electronic devices during the session

Verify this list with the examination team either when you register or before the testing session begins to ensure the session has no other restrictions. See `http://www.arrl.org/what-to-bring-to-an-exam-session` for more information about the exam session.

You will have all the time necessary to complete the exam. Do not rush. Read each question carefully and be sure to indicate the correct answer. There is no penalty for guessing. If you must guess the answer, try to eliminate as many choices as possible for the question. Then select the option that seems the most correct to you.

Good luck with the exam. I hope to meet you on the bands one day!

Chapter 1

G1 — COMMISSION'S RULES

1.1 Introduction

As we did in the Technician Class license study guide, we begin our General Class license study with rules and regulations that govern the Amateur Radio Service. Because the General Class license grants more operating frequency privileges with worldwide access than does the Technician Class license, there will be more detailed questions on permitted operations. The questions will frequently refer to the Federal Communications Commission (FCC) rules in the Code of Federal Regulations (CFR). Part 97, the Amateur Radio Service, is in Chapter I, Subchapter D of Title 47. You can find the current text for Part 97 by following the links in `https://www.ecfr.gov/cgi-bin/text-idx?c=ecfr&tp l=/ecfrbrowse/Title47/47cfrv5_02.tpl`.

As a review, we must remember that the states in the United States (US) are in International Telecommunication Union (ITU) Region 2 (see Figure 1.1). *Note:* not all US territories are in Region 2. We have special agreements within Region 2 with our immediate neighbors in Canada and Mexico, as you might expect. The URL for the ITU is `http://www.itu.int/en/Pages/default.aspx`. A final Web site you may find useful is for the International Amateur Radio Union (IARU) `http://www.iaru.org`.

As we look at Part 97 in more detail, you may wish to make study cards to remember all the frequency divisions that the exam may have. As we look at the questions and answers in this chapter, we will see quotes from Part 97 that are relevant. Next to the questions, the question pool also lists the specific section in Part 97 where the question originates. This Part 97 information will not appear on the exam questions – it is only for study purposes.

The *Commission's Rules* subelement has the following question groups:

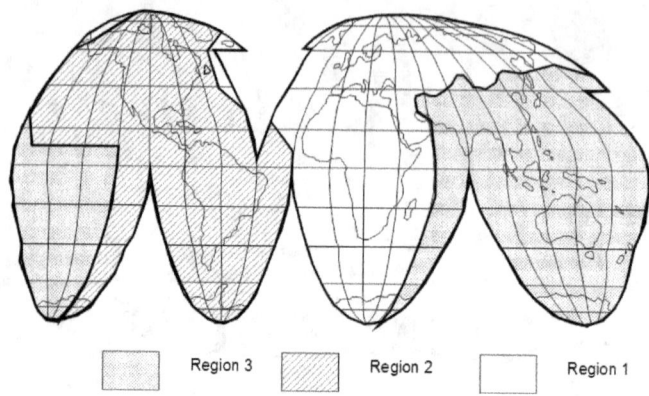

| | Region 3 | | Region 2 | | Region 1 |

Figure 1.1: The ITU regions around the world. The U.S. is in ITU Region 2.

A. Operator Frequency Privileges
B. Amateur Practice
C. Emissions
D. Licensing
E. Station Control

Subelement 1 will generate five questions on the General Class examination.

1.2   Radio Engineering Concepts

Frequency Bands   The General Class operator's frequency band access is more complicated than we saw for Technician Class operators. Table 1.1 summarizes the General Class privileges on the Low Frequency (LF), Medium Frequency (MF), and High Frequency (HF) bands. See part 97 for the full list of privileges on all bands. Note: Continuous Wave (CW) is allowed on all bands, so the table does not explicitly list it.

Emission Types   There are several questions where the exam may ask you about specific emission types. We saw these in the Technician Class license study, and we need to remember them again for the General Class license exam. The Part 97 definitions for the emission types are

CW — International Morse code telegraphy emissions
Data — Telemetry, telecommand and computer communications emissions
Image — Facsimile and television emissions
MCW — Tone-modulated international Morse code telegraphy emissions
Phone — Speech and other sound emissions
RTTY — Narrow-band direct-printing telegraphy emissions
SS — Spread spectrum emissions using bandwidth-expansion modulation
Test — Emissions containing no information

Table 1.1: Non-CW General Class Emission Permissions in Region 2.

Band (m)	Permissions	Emissions
2200	135.7–137.8 kHz	RTTY, data, phone, image
630	472–479 kHz	RTTY, data, phone, image
160	1800–2000 kHz	RTTY, data, phone, image
80	3.525–3.600 MHz	RTTY, data
75	3.800–4.000 MHz	phone, image
60	5.3305–5.4064 MHz	phone, RTTY, data
40	7.025–7.125 MHz	RTTY, data
	7.175–7.300 MHz	phone, image
30	10.100–10.150 MHz	RTTY, data
20	14.025–14.150 MHz	RTTY, data
	14.225–14.350 MHz	phone, image
17	18.068–18.110 MHz	RTTY, data
	18.110–18.168 MHz	phone, image
15	21.025–21.200 MHz	RTTY, data
	21.275–21.450 MHz	phone, image
12	24.890–24.930 MHz	RTTY, data
	24.930–24.990 MHz	phone, image
10	28.000–28.300 MHz	RTTY, data
	28.300–29.700 MHz	phone, image

Allowed Data Rates When an operator transmits with digital data formats, the required bandwidth scales with the data transmission rate. Engineers determine the digital transmission's bandwidth is by how frequently the modulation changes state. The modulation's rate of change per second is called the *baud rate* with the unit of Symbols per second (sps). Currently-popular digital modulation techniques on amateur bands generally send one bit with each modulation state. For example, with PSK31, the modulation uses Binary Phase Shift Keying (BPSK) for most transmissions, and BPSK sends one bit per modulation state. Therefore the bit rate of 31 bps is a baud rate of 31 baud. However, PSK31 software also has a Quadrature Phase Shift Keying (QPSK) mode that sends two bits in each modulation state. In this mode, PSK31 still transmits at 31 baud, the modulation symbol rate, even though it is transmitting a source data rate of 62 bps. Knowing the transmission baud is important because, as a rule of thumb, the required transmission bandwidth is approximately 1 Hz of bandwidth for every 1 baud transmission rate. As we can see in Table 1.1, each band has a different bandwidth, and some bands have sub-bands where you can find the digital data. Table 1.2 lists the allowable baud rates for different bands.

Table 1.2: Allowed Baud Rates on the Amateur Bands.

Band (m)	Allowed Rate
160	300 baud
80	300 baud
40	300 baud
30	300 baud
20	300 baud
17	300 baud
15	300 baud
12	300 baud
10	1200 baud
6	19.6 kilobaud
2	19.6 kilobaud
1.25	56 kilobaud
0.70	56 kilobaud

1.3 G1A - Operator Frequency Privileges

1.3.1 Overview

The *Operator Frequency Privileges* question group in Subelement G1 tests you on the allowed General Class MF and HF operating bands, and the emissions permitted on those bands. The *Operator Frequency Privileges* group covers topics such as

- General Class control operator frequency privileges
- Primary and secondary allocations

The test producer will select one of the 15 questions in this group for your exam.

1.3.2 Questions

G1A01 [97.301(d)] On which HF/MF bands is a General Class license holder granted all amateur frequency privileges?
   A. 60, 20, 17, and 12 meters
   B. 160, 80, 40, and 10 meters
   C. 160, 60, 30, 17, 12, and 10 meters
   D. 160, 30, 17, 15, 12, and 10 meters

If you go to Part 97 or Table 1.1, you will see that a General Class operator has frequency privileges on 2200, 630, 160, 80/75, 60, 40, 30, 20, 17, 15, 12, and 10 meters. However, if you look closely, you will see that General Class amateurs have full access across the entire band on the 160, 60, 30, 17, 12, and 10 meters of the bands in this question. This access makes **Answer C** the correct choice. The other choices include bands where General Class amateurs have some, but not full, access across the entire band.

G1A02 [97.305] On which of the following bands is phone operation prohibited?
A. 160 meters
B. 30 meters
C. 17 meters
D. 12 meters

If you inspect Part 97 or Table 1.1, you will see that the regulations only permit Radio TeleType (RTTY) and data on the 30-m band. Since we are looking for a band that does not allow phone, **Answer B** is the right choice. The other bands permit phone transmissions.

G1A03 [97.305] On which of the following bands is image transmission prohibited?
A. 160 meters
B. 30 meters
C. 20 meters
D. 12 meters

From the last question, you should be able to spot the correct answer here. We saw earlier that the 30-m band only allows RTTY and data. It does not allow image transmission, so the correct choice is **Answer B**. From Part 97 or Table 1.1, we know that the other bands in this question allow image data.

G1A04 [97.303 (h)] Which of the following amateur bands is restricted to communication on only specific channels, rather than frequency ranges?
A. 11 meters
B. 12 meters
C. 30 meters
D. 60 meters

Part 97 states that in "the 5330.5-5406.4 kHz band (60 m band), amateur stations may transmit only on the five center frequencies specified.... [a]mateur operators shall ensure that their emissions do not occupy more than 2.8 kHz centered on each of these center frequencies." The 60-m band is the one channelized, so **Answer D** is the correct choice.

G1A05 [97.301(d)] Which of the following frequencies is in the General class portion of the 40-meter band (in ITU Region 2)?
A. 7.250 MHz
B. 7.500 MHz
C. 40.200 MHz
D. 40.500 MHz

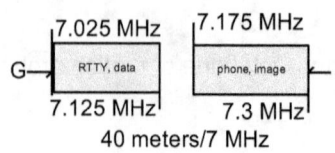

Figure 1.2: The 40-m/7-MHz band.

Figure 1.2 shows the 40-m band runs from 7.0 MHz to 7.3 MHz. The General Class sub-band runs from 7.025 MHz to 7.125 MHz and from 7.175 MHz to 7.3 MHz. Of the choices given, only **Answer A**, 7.250 MHz, is within the General Class operator's range. Answers C and D are not on the 40-m band.

G1A06 [97.301(d)] Which of the following frequencies is within the General class portion of the 75-meter phone band?
A. 1875 kHz
B. 3750 kHz
C. 3900 kHz
D. 4005 kHz

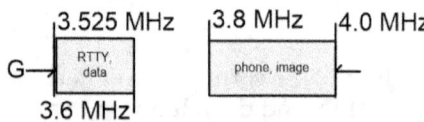

Figure 1.3: The 75 & 80-m/3.5-MHz band.

Figure 1.3 illustrates the 75-m and 80-m bands. The 75-m phone sub-band for General Class operators runs from 3800 kHz to 4000 kHz. 3750 kHz is in the Extra Class sub-band, not in the General Class sub-band, so Answer B is incorrect. **Answer C** is correct because 3900 kHz is in the General Class phone sub-band.

G1A07 [97.301(d)] Which of the following frequencies is within the General class portion of the 20-meter phone band?
A. 14005 kHz
B. 14105 kHz
C. 14305 kHz
D. 14405 kHz

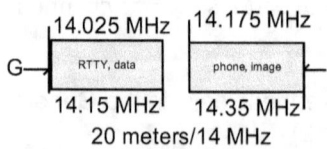

Figure 1.4: The 20-m/14-MHz band.

The reasoning here is like that in the previous question: the General Class portion of the 20-m band is from 14.025 MHz to 14.15 MHz and from 14.175 MHz to 14.350 MHz. The 14305 kHz of **Answer C** is the only choice given in this range. The other options are outside this region.

G1A08 [97.301(d)] Which of the following frequencies is within the General class portion of the 80-meter band?
  A. 1855 kHz
  B. 2560 kHz
  C. 3560 kHz
  D. 3650 kHz

As Figure 1.3 shows, the General Class sub-band of the 80-m band runs from 3525 kHz to 3600 kHz. This makes **Answer C**, 3560 kHz, the right choice. Be careful with Answer D because it is in the Extra class sub-band, not the General Class sub-band. The 1855 kHz in Answer A is on the 160-m band.

G1A09 [97.301(d)] Which of the following frequencies is within the General class portion of the 15-meter band?
  A. 14250 kHz
  B. 18155 kHz
  C. 21300 kHz
  D. 24900 kHz

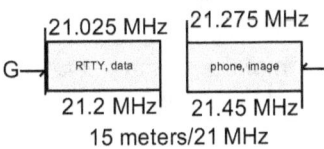

15 meters/21 MHz

By looking at Figure 1.5, you should be able to spot **Answer C**, 21 300 kHz, as the right choice for this question. Answer A is on the 20-m band, Answer B is on the 17-m band, while Answer D is on the 12-m band.

Figure 1.5: The 15-m/21-MHz band.

G1A10 [97.301(d)] Which of the following frequencies is available to a control operator holding a General class license?
  A. 28.020 MHz
  B. 28.350 MHz
  C. 28.550 MHz
  D. All of these choices are correct

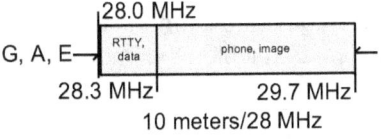

10 meters/28 MHz

If we look at Figure 1.6 for the 10-m band, we can see that each of the frequencies in Answers A, B, and C are available to General Class operators, which makes **Answer D** the correct choice.

Figure 1.6: The 10-m/28-MHz band.

G1A11 [97.301] When General class licensees are not permitted to use the entire voice portion of a band, which portion of the voice segment is generally available to them?
- A. The lower frequency end
- B. The upper frequency end
- C. The lower frequency end on frequencies below 7.3 MHz, and the upper end on frequencies above 14.150 MHz
- D. The upper frequency end on frequencies below 7.3 MHz, and the lower end on frequencies above 14.150 MHz

If you inspect the band graphics on the previous questions, you will see that when Part 97 restricts the General Class operator on the band, the allowed frequencies for phone emissions are at the upper end of the band. This restriction makes **Answer B** the right choice. Be careful with Answers C and D because they are trying to sound like the convention on Upper Side Band (USB) vs. Lower Side Band (LSB) for Single Sideband (SSB) operations. Answer A is the opposite of the restriction.

G1A12 [97.303] Which of the following applies when the FCC rules designate the Amateur Service as a secondary user on a band?
- A. Amateur stations must record the call sign of the primary service station before operating on a frequency assigned to that station
- B. Amateur stations can use the band only during emergencies
- C. Amateur stations can use the band only if they do not cause harmful interference to primary users
- D. Amateur stations may only operate during specific hours of the day, while primary users are permitted 24-hour use of the band

When Part 97 designates a service as a secondary user of a band, then "station in a secondary service must not cause harmful interference to, and must accept interference from, stations in a primary service." **Answer C** the only choice that is consistent with Part 97.

G1A13 [97.303(h)(2)(j)] What is the appropriate action if, when operating on either the 30-meter or 60-meter bands, a station in the primary service interferes with your contact?
- A. Notify the FCCs regional Engineer in Charge of the interference
- B. Increase your transmitter's power to overcome the interference
- C. Attempt to contact the station and request that it stop the interference
- D. Move to a clear frequency or stop transmitting

As we saw in the previous question, the secondary user must accept the interference. The only way to get around this is to either move to a clear frequency or to stop transmitting, as in **Answer D**. The other choices are not consistent with Part 97 and good operating practice.

G1A14 [97.301(d)] Which of the following may apply in areas under FCC jurisdiction outside of ITU Region 2?
  A. Station identification may have to be in a language other than English
  B. Morse code may not be permitted
  C. Digital transmission may not be permitted
  D. Frequency allocations may differ

When we go to Part 97, we see that the permitted emission types outside of ITU Region 2 are the same as those inside Region 2, so Answers B and C are incorrect. Since the station is still under the FCC's domain, the identification must be in English, which makes Answer A incorrect. However, Part 97 does show that some of the operating frequencies are different outside of Region 2, which makes **Answer D** the correct choice.

G1A15 [97.205(b)] What portion of the 10-meter band is available for repeater use?
  A. The entire band
  B. The portion between 28.1 MHz and 28.2 MHz
  C. The portion between 28.3 MHz and 28.5 MHz
  D. The portion above 29.5 MHz

For a 10-m repeater, Part 97 states that a "repeater may receive and retransmit ... except the 28.0-29.5 MHz" segment. This restriction makes Answer A incorrect. Answers B and C are in the non-allowed segment, so they are incorrect. **Answer D** uses an allowed frequency, so this is the right answer.

1.4 G1B - Amateur Practice

1.4.1 Overview

The *Amateur Practice* question group in Subelement G1 covers a variety of topics from Part 97 relating to station placement and operations. The *Amateur Practice* group covers topics such as
  • Antenna structure limitations
  • Good engineering and good amateur practice
  • Beacon operation
  • Prohibited transmissions
  • Retransmitting radio signals
The test producer will select one of the 12 questions in this group for your exam.

1.4.2 Questions

G1B01 [97.15(a)] What is the maximum height above ground to which an antenna structure may be erected without requiring notification to the FAA and registration with the FCC, provided it is not at or near a public use airport?
- A. 50 feet
- B. 100 feet
- C. 200 feet
- D. 300 feet

The Part 97 rule is that owners "of certain antenna structures more than 60.96 meters (200 feet) above ground level at the site or located near or at a public use airport must notify the Federal Aviation Administration and register with the Commission." This rule makes **Answer C** the correct choice.

G1B02 [97.203(b)] With which of the following conditions must beacon stations comply?
- A. A beacon station may not use automatic control
- B. The frequency must be coordinated with the National Beacon Organization
- C. The frequency must be posted on the Internet or published in a national periodical
- D. There must be no more than one beacon signal transmitting in the same band from the same station location

The rule in Part 97 states that a "beacon must not concurrently transmit on more than 1 channel in the same amateur service frequency band, from the same station location." Only **Answer D** meets the regulation, so it is the right choice.

G1B03 [97.3(a)(9)] Which of the following is a purpose of a beacon station as identified in the FCC rules?
- A. Observation of propagation and reception
- B. Automatic identification of repeaters
- C. Transmission of bulletins of general interest to Amateur Radio licensees
- D. Identifying net frequencies

Part 97 defines a beacon station as an "amateur station transmitting communications for the purposes of observation of propagation and reception or other related experimental activities." Only **Answer A** matches the definition.

G1B04 [97.113(c)] Which of the following transmissions is permitted?
- A. Unidentified transmissions for test purposes only
- B. Retransmission of other amateur station signals by any amateur station
- C. Occasional retransmission of weather and propagation forecast information from U.S. government stations
- D. Coded messages of any kind, if not intended to facilitate a criminal act

The regulation in Part 97 states that

> No station shall retransmit programs or signals emanating from any type of radio station other than an amateur station, except propagation and weather forecast information intended for use by the general public and originated from United States Government stations, and communications, including incidental music, originating on United States Government frequencies between a manned spacecraft and its associated Earth stations.

As can be seen, **Answer C** fits this criterion of Part 97. The other choices are not allowed under Part 97.

G1B05 [97.111((5)(b)] Which of the following one-way transmissions are permitted?
  A. Unidentified test transmissions of less than one minute in duration
  B. Transmissions necessary to assist learning the International Morse code
  C. Regular transmissions offering equipment for sale, if intended for Amateur Radio use
  D. All these choices are correct

Part 97 permits one-way communications "to assisting persons learning, or improving proficiency in, the international Morse code". Only **Answer B** is correct among the choices given.

G1B06 [97.15(b), PRB-1, 101 FCC 2d 952 (1985)] Under what conditions are state and local governments permitted to regulate Amateur Radio antenna structures?
  A. Under no circumstances, FCC rules take priority
  B. At any time and to any extent necessary to accomplish a legitimate purpose of the state or local entity, provided that proper filings are made with the FCC
  C. Only when such structures exceed 50 feet in height and are clearly visible 1000 feet from the structure
  D. Amateur Service communications must be reasonably accommodated, and regulations must constitute the minimum practical to accommodate a legitimate purpose of the state or local entity

This question addresses the balancing of local governance with a defined activity permitted under national authority. Part 97 states that "[s]tate and local regulation of a station antenna structure must not preclude amateur service communications. Rather, it must reasonably accommodate such communications and must constitute the minimum practicable regulation to accomplish the state or local authority's legitimate purpose." **Answer D** is the only choice matching this wording.

G1B07 [97.113(a)(4)] What are the restrictions on the use of abbreviations or procedural signals in the Amateur Service?
A. Only "Q" signals are permitted
B. They may be used if they do not obscure the meaning of a message
C. They are not permitted
D. Only "10 codes" are permitted

While coded messages are generally not permitted with the Amateur Service, standard abbreviations, procedural signs, and "Q" signals are not coded transmissions used to disguise the meaning of the transmission. Instead, they are well-known communications shorthand expressions commonly accepted in the community and do not obscure the operator's purpose when used in a commonly-accepted manner. Therefore, **Answer B** is the correct response.

G1B08 [97.101(a)] When choosing a transmitting frequency, what should you do to comply with good amateur practice?
A. Insure that the frequency and mode selected are within your license class privileges
B. Follow generally accepted band plans agreed to by the Amateur Radio community
C. Monitor the frequency before transmitting
D. All of these choices are correct

Each of the steps identified in Answers A, B, and C is part of good amateur practice, so the best choice is **Answer D**.

G1B09 [97.203(d)] On what HF frequencies are automatically controlled beacons permitted?
A. On any frequency if power is less than 1 watt
B. On any frequency if transmissions are in Morse code
C. 21.08 MHz to 21.09 MHz
D. 28.20 MHz to 28.30 MHz

The regulation in Part 97 states that a "beacon may be automatically controlled while it is transmitting on the 28.20-28.30 MHz" band segment. Only **Answer D** matches Part 97.

G1B10 [97.203(c)] What is the power limit for beacon stations?
A. 10 watts PEP output
B. 20 watts PEP output
C. 100 watts PEP output
D. 200 watts PEP output

The Part 97 rules on beacon stations state that the "transmitter power of a beacon must not exceed 100 W." The maximum power level of **Answer C** matches

Part 97.

G1B11 [97.101(a)] Who or what determines "good engineering and good amateur practice," as applied to the operation of an amateur station in all respects not covered by the Part 97 rules?
  A. The FCC
  B. The control operator
  C. The IEEE
  D. The ITU

As in **Answer A**, the FCC regulates amateur radio practices in the US, so that agency makes the final determination on good practices. The ITU does not regulate in the US, making this an incorrect choice. The Institute of Electrical and Electronics Engineers (IEEE) is a professional group, and it has no regulatory function. Answer B would allow the operator to set standards, so that is not a good choice.

G1B12 [97.111(a)(1)] When is it permissible to communicate with amateur stations in countries outside the areas administered by the Federal Communications Commission?
  A. Only when the foreign country has a formal third-party agreement filed with the FCC
  B. When the contact is with amateurs in any country except those whose administrations have notified the ITU that they object to such communications
  C. When the contact is with amateurs in any country as long as the communication is conducted in English
  D. Only when the foreign country is a member of the International Amateur Radio Union

Part 97 permits operators to send "transmissions necessary to exchange messages with other stations in the amateur service, except those in any country whose administration has notified the ITU that it objects to such communications. The FCC will issue public notices of current arrangements for international communications." This rule means we can contact amateurs in other countries except for those countries that have notified the ITU that such contacts are not welcome. Therefore, **Answer B** is the correct choice.

1.5 G1C - Emissions

1.5.1 Overview

The *Emissions* question group in Subelement G1 quizzes you on allowed power levels, emission types, and data rates on the Amateur bands. The *Emissions* group covers topics such as
- Transmitter power regulations
- Data emission standards
- 60-meter operation requirements

The test producer will select one of the 15 questions in this group for your exam.

1.5.2 Questions

G1C01 [97.313(c)(1)] What is the maximum transmitting power an amateur station may use on 10.140 MHz?
- A. 200 watts PEP output
- B. 1000 watts PEP output
- C. 1500 watts PEP output
- D. 2000 watts PEP output

For this question, and the following ones on power limits, we need to remember one fact: except for Spread Spectrum (SS) communications, the FCC maximum legal power limits are either 1500 W or 200 W under normal operating conditions. Generally, the 200-W limit is on the Novice/Tech+ parts of the radio spectrum below 28.1 MHz, while on the other bands, except for 30 m and 60 m, we can use up to 1500 W. All other power choices are to distract you and can be ignored. The correct answer here comes from either Answer A or Answer C. The 30-m band is an exception, and everyone is limited to a maximum power of 200 W. Therefore, **Answer A** is the correct answer.

G1C02 [97.313] What is the maximum transmitting power an amateur station may use on the 12-meter band?
- A. 50 watts PEP output
- B. 200 watts PEP output
- C. 1500 watts PEP output
- D. An effective radiated power equivalent to 100 watts from a half-wave dipole

Since there are no specific restrictions in Part 97, one may use up to 1500 W Peak Envelope Power (PEP), and **Answer C** is the correct choice.

G1C03 [97.303(h)(1)] What is the maximum bandwidth permitted by FCC rules for Amateur Radio stations transmitting on USB frequencies in the 60-meter band?
  A. 2.8 kHz
  B. 5.6 kHz
  C. 1.8 kHz
  D. 3 kHz

The Part 97 rules in this case state that "[a]mateur operators shall ensure that their transmission occupies only the 2.8 kHz centered around each of these frequencies." Only **Answer A** matches Part 97.

G1C04 [97.313(a)] Which of the following limitations apply to transmitter power on every amateur band?
  A. Only the minimum power necessary to carry out the desired communications should be used
  B. Power must be limited to 200 watts when using data transmissions
  C. Power should be limited as necessary to avoid interference to another radio service on the frequency
  D. Effective radiated power cannot exceed 1500 watts

Even though we may legally be permitted to use up to 1500 W for transmissions, both good practice and the Part 97 regulations state that an "amateur station must use the minimum transmitter power necessary to carry out the desired communications." This principle is found in **Answer A**, making it correct.

G1C05 [97.313] What is the limit for transmitter power on the 28 MHz band for a General Class control operator?
  A. 100 watts PEP output
  B. 1000 watts PEP output
  C. 1500 watts PEP output
  D. 2000 watts PEP output

According to Part 97, we can legally operate at a maximum level of 1500 W PEP, and **Answer C** is the correct answer.

G1C06 [97.313] What is the limit for transmitter power on the 1.8 MHz band?
  A. 200 watts PEP output
  B. 1000 watts PEP output
  C. 1200 watts PEP output
  D. 1500 watts PEP output

As we have noted, the correct answer here comes from either Answer A or Answer D. Part 97 says that we can use up to the full legal limit of 1500 W PEP with the reminder that good operating practice says we should never use more power than necessary. Therefore, **Answer D** is the correct choice.

G1C07 [97.305(c), 97.307(f)(3)] What is the maximum symbol rate permitted for RTTY or data emission transmission on the 20-meter band?
A. 56 kilobaud
B. 19.6 kilobaud
C. 1200 baud
D. 300 baud

Digital transmissions like RTTY occupy about 1 Hz of bandwidth for every baud of data rate. As we saw in Table 1.2, Part 97 specifies that on the 20-m band, "the symbol rate must not exceed 300 bauds" so **Answer D** is correct. Answer A is for above 2 m, Answer B is for above 28 MHz (above the 10-m band), and Answer C is for on the 10-m band.

G1C08 [97.307(f)(3)] What is the maximum symbol rate permitted for RTTY or data emission transmitted at frequencies below 28 MHz?
A. 56 kilobaud
B. 19.6 kilobaud
C. 1200 baud
D. 300 baud

Part 97 states that one can use 1200 baud on 10 m/28 MHz. However, this question is asking for the rate at frequencies below 28 MHz, so the 300-baud restriction is still in place, as it was in the previous question (see also Table 1.2). **Answer D** is correct for this question.

G1C09 [97.305(c) and 97.307(f)(5)] What is the maximum symbol rate permitted for RTTY or data emission transmitted on the 1.25-meter and 70-centimeter bands?
A. 56 kilobaud
B. 19.6 kilobaud
C. 1200 baud
D. 300 baud

Part 97 states that the "symbol rate must not exceed 56 kilobauds" for these bands. This rate makes **Answer A** the correct choice. Answer D is for below 10-m use, Answer C is for use on 10 m, and Answer B is for 2 or 6-m use.

G1C10 [97.305(c) and 97.307(f)(4)] What is the maximum symbol rate permitted for RTTY or data emission transmissions on the 10-meter band?
A. 56 kilobaud
B. 19.6 kilobaud
C. 1200 baud
D. 300 baud

Now we have moved to the 10-m band. Here, the FCC stipulates that the "symbol rate must not exceed 1200 bauds," and the correct choice is **Answer C**. Read the question carefully because previous questions were concerned about operating below 10 m, and this question is for using 10 m, so Answer D is wrong. Answer A is for use above 2 m, while Answer B is for use at 2 m and 6 m. Look at Table 1.2 again if you need to.

G1C11 [97.305(c) and 97.307(f)(5)] What is the maximum symbol rate permitted for RTTY or data emission transmissions on the 2-meter band?
  A. 56 kilobaud
  B. 19.6 kilobaud
  C. 1200 baud
  D. 300 baud

Here we are at 2 m. As we can see in Table 1.2, we can run at 19.6 kilobaud, which makes **Answer B** correct. Answer C is for use on the 10-m band, while Answer D is for use below 10 m. Answer A is for use on the bands above 2 m.

G1C12 [97.313(i)] Which of the following is required by the FCC rules when operating in the 60-meter band?
  A. If you are using an antenna other than a dipole, you must keep a record of the gain of your antenna
  B. You must keep a record of the date, time, frequency, power level, and stations worked
  C. You must keep a record of all third-party traffic
  D. You must keep a record of the manufacturer of your equipment and the antenna used

Part 97 states that for 60-m operations

> No station may transmit with an effective radiated power (ERP) exceeding 100 W PEP on the 60 m band. For the purpose of computing ERP, the transmitter PEP will be multiplied by the antenna gain relative to a half-wave dipole antenna. A half-wave dipole antenna will be presumed to have a gain of 1 (0 dBd). Licensees using other antennas must maintain in their station records either the antenna manufacturer's data on the antenna gain or calculations of the antenna gain.

Answer A is a description of one of the transmitting antenna requirements for 60 m. The other choices are to distract you.

G1C13 [97.309(a)(4)] What must be done before using a new digital protocol on the air?
   A. Type-certify equipment to FCC standards
   B. Obtain an experimental license from the FCC
   C. Publicly document the technical characteristics of the protocol
   D. Submit a rule-making proposal to the FCC describing the codes and methods of the technique

Part 97 permits amateurs to develop and use new digital transmission protocols if the "amateur station transmitting a RTTY or data emission using a digital code specified in this paragraph may use any technique whose technical characteristics have been documented publicly". **Answer C** is the correct choice.

G1C14 [97.313(i)] What is the maximum power limit on the 60-meter band?
   A. 1500 watts PEP
   B. 10 watts RMS
   C. ERP of 100 watts PEP with respect to a dipole
   D. ERP of 100 watts PEP with respect to an isotropic antenna

As we saw above, an amateur station can use a maximum transmission power of 100 W relative to a dipole, so **Answer C** is the right choice. Be careful with Answer D because it has the right power and the wrong reference antenna.

G1C15 [97.313] What measurement is specified by FCC rules that regulate maximum power output?
   A. RMS
   B. Average
   C. Forward
   D. PEP

As you look through Part 97, you will see that it lists the maximum power limits in terms of the PEP. This wording makes **Answer D** the correct choice. We will see the differences between Root Mean Square (RMS) and PEP again in Chapter 5 when we discuss electrical quantities.

1.6   G1D - Licensing

1.6.1   Overview

The *Licensing* question group in Subelement G1 tests you over procedures for Amateur licensing and Volunteer Examiners. The *Licensing* group covers topics such as
   • Volunteer Examiners and Volunteer Examiner Coordinators
   • Temporary identification

• Element credit

The test producer will select one of the 11 questions in this group for your exam.

1.6.2 Questions

G1D01 [97.501, 97.505(a)] Who may receive partial credit for the elements represented by an expired Amateur Radio license?
- A. Any person who can demonstrate that they once held an FCC-issued General, Advanced, or Amateur Extra class license that was not revoked by the FCC
- B. Anyone who held an FCC-issued Amateur Radio license that has been expired for not less than 5 years and not more than 15 years
- C. Any person who previously held an amateur license issued by another country, but only if that country has a current reciprocal licensing agreement with the FCC
- D. Only persons who once held an FCC issued Novice, Technician, or Technician Plus license

If the holder of an expired General, Advanced, or Amateur Extra license was in good standing with the FCC, then Part 97 mandates that the holder has partial credit for that expired license. This rule makes **Answer A** the right choice.

G1D02 [97.509(b)(3)(i)] What license examinations may you administer when you are an accredited VE holding a General class operator license?
- A. General and Technician
- B. General only
- C. Technician only
- D. Extra, General and Technician

In Part 97, we see that a Volunteer Examiner (VE) must hold an "Amateur Extra, Advanced or General Class in order to administer a Technician Class operator license examination." It is the only one that a General Class operator can prepare, so **Answer C** is the correct choice.

G1D03 [97.9(b)] On which of the following band segments may you operate if you are a Technician class operator and have a Certificate of Successful Completion of Examination (CSCE) for General class privileges?
- A. Only the Technician band segments until your upgrade is posted in the FCC database
- B. Only on the Technician band segments until your license arrives in the mail
- C. On any General or Technician class band segment
- D. On any General or Technician class band segment except 30-meters and 60-meters

The general rule in Part 97 for a licensed operator holding a Certificate of Successful Completion of an Examination (CSCE) is that operator "is authorized to exercise the rights and privileges of the higher operator class until final disposition of the application or until 365 days following the passing of the examination, whichever comes first." This rule means that the operator has access to the General Class band segments in addition to the existing Technician Class privileges, as in **Answer C**. The other choices do not match the regulations.

G1D04 [97.509(3)(i)(c)] Which of the following is a requirement for administering a Technician class license examination?
  A. At least three General class or higher VEs must observe the examination
  B. At least two General class or higher VEs must be present
  C. At least two General class or higher VEs must be present, but only one need be Amateur Extra class
  D. At least three VEs of Technician class or higher must observe the examination

The FCC requires that each "examination for an amateur operator license must be administered by a team of at least 3 VEs at an examination session coordinated by a VEC." The minimum license grade to be a VE is the General Class, so the correct choice is **Answer A**. Answer B is wrong because the minimum number of VEs is incorrect. Answer C is wrong because the number of VEs is incorrect, and an Extra Class VE is not required. Answer D is wrong because Technician Class operators cannot be VEs.

G1D05 [97.509(b)(3)(i)] Which of the following must a person have before they can be an administering VE for a Technician class license examination?
  A. Notification to the FCC that you want to give an examination
  B. Receipt of a Certificate of Successful Completion of Examination (CSCE) for General class
  C. Possession of a properly obtained telegraphy license
  D. An FCC General class or higher license and VEC accreditation

To be eligible to participate as a VE, the operator must have at least a General Class license in hand from the FCC and a valid accreditation from a Volunteer Examiner Coordinator (VEC), so the correct choice is **Answer D**. Although it may sound logical, the FCC is no longer directly involved with administering the exams, so Answer A is out. Answer B cannot be correct because the FCC requires that the VE has their license grant, while the CSCE is a temporary permit until the paperwork is validated and finalized, and the license grant is issued. Answer C is wrong because there is no amateur telegraphy license.

G1D06 [97.119(f)(2)] When must you add the special identifier "AG" after your call sign if you are a Technician class licensee and have a Certificate of Successful Completion of Examination (CSCE) for General class operator privileges, but the FCC has not yet posted your upgrade on its website?
  A. Whenever you operate using General class frequency privileges
  B. Whenever you operate on any amateur frequency
  C. Whenever you operate using Technician frequency privileges
  D. A special identifier is not required as long as your General class license application has been filed with the FCC

You must use the identifier with your new privileges until your license grant has been fully processed and entered into the FCC database. Answers B and C may sound correct, but you do not need the "AG" identification if you operate with your existing Technician Class privilege frequencies, so they are wrong. Answer D may sound correct, but the FCC finalizes the license grant when they enter the information into their database and not when they first receive it for processing. The correct procedure is when you are using General Class privileges, as in **Answer A.**

G1D07 [97.509(b)(1)] Volunteer Examiners are accredited by what organization?
  A. The Federal Communications Commission
  B. The Universal Licensing System
  C. A Volunteer Examiner Coordinator
  D. The Wireless Telecommunications Bureau

Looking at Part 97, we see that each "administering VE must ... [b]e accredited by the coordinating VEC." This wording makes **Answer C** the right choice. The FCC does not certify VEs, so Answers A and D are incorrect. The Universal Licensing System (ULS) stores license information, but it cannot approve a VE, so Answer B is also wrong.

G1D08 [97.509(b)(3)] Which of the following criteria must be met for a non-U.S. citizen to be an accredited Volunteer Examiner?
  A. The person must be a resident of the U.S. for a minimum of 5 years
  B. The person must hold an FCC granted Amateur Radio license of General Class or above
  C. The person's home citizenship must be in ITU region 2
  D. None of these choices is correct; a non-U.S. citizen cannot be a Volunteer Examiner

The FCC requires that a VE be "a person who holds an amateur operator license of the class ... Amateur Extra, Advanced or General Class", so **Answer B** is the right choice. Since there is no citizenship or residency requirement, Answers A, C, and D are incorrect.

G1D09 [97.9(b)] How long is a Certificate of Successful Completion of Examination (CSCE) valid for exam element credit?

A. 30 days
B. 180 days
C. 365 days
D. For as long as your current license is valid

The CSCE shows each examination element "the examinee passed within the previous 365 days." **Answer C** is the only option matching Part 97.

G1D10 [97.509(b)(2)] What is the minimum age that one must be to qualify as an accredited Volunteer Examiner?

A. 12 years
B. 18 years
C. 21 years
D. There is no age limit

Part 97 specifies that a VE must be "at least 18 years of age." This rule makes **Answer B** the correct choice.

G1D11 [97.505] What is required to obtain a new General Class license after a previously-held license has expired and the two-year grace period has passed?

A. They must have a letter from the FCC showing they once held an amateur or commercial license
B. There are no requirements other than being able to show a copy of the expired license
C. The applicant must be able to produce a copy of a page from a call book published in the U.S. showing his or her name and address
D. The applicant must pass the current element 2 exam

Part 97 indicates that license holders in this situation can receive credit for Element 3 or 4, depending upon the license class. However, they do not obtain credit for Element 2 (the Technician Class license). Therefore, to obtain a new license in compliance with Part 97, they must pass the Element 2 examination, as in **Answer D**.

1.7   G1E - Station Control

1.7.1   Overview

The *Station Control* question group in Subelement G1 quizzes you on Part 97 regulations dealing with being a control operator in various situations. The *Station Control* group covers topics such as

- Control categories

- Repeater regulations
- Third-party rules
- ITU regions
- Automatically controlled digital station

The test producer will select one of the 11 questions in this group for your exam.

1.7.2 Questions

G1E01 [97.115(b)(2)] Which of the following would disqualify a third party from participating in stating a message over an amateur station?
   A. The third party's amateur license has been revoked and not reinstated
   B. The third party is not a U.S. citizen
   C. The third party is a licensed amateur
   D. The third party is speaking in a language other than English

The section in the Part 97 regulations that is the subject of this question states that the

> third party is not a prior amateur service licensee whose license was revoked; suspended for less than the balance of the license term and the suspension is still in effect; suspended for the balance of the license term and relicensing has not taken place; or surrendered for cancellation following notice of revocation, suspension or monetary forfeiture proceedings. The third party may not be the subject of a cease and desist order which relates to amateur service operation and which is still in effect.

Notice that while the license must not have been revoked or suspended, there is no nationality, current amateur license, or language restriction in the rule. Therefore, **Answer A** is the right choice.

G1E02 [97.205(b)] When may a 10-meter repeater retransmit the 2-meter signal from a station having a Technician class control operator?
   A. Under no circumstances
   B. Only if the station on 10-meters is operating under a Special Temporary Authorization allowing such retransmission
   C. Only during an FCC declared general state of communications emergency
   D. Only if the 10-meter repeater control operator holds at least a General class license

Believe it or not, a Technician class operator can use Frequency Modulation (FM) phone on 10 m if they do it the right way, so Answer A is incorrect. Part 97 states that a "holder of a Technician, General, Advanced or Amateur Extra class operator license may be the control operator of a repeater, subject to the privileges of the class of operator license held." Therefore, each transmitter may only operate on those frequencies where the control operator has a license grant.

Every time a new repeater passes a message, it is "re-born," and it falls under the license of the repeater's control operator. Therefore, if the repeater's control operator has at least General Class privileges, the repeater can retransmit the 2-m signal on the 10-m band, and **Answer D** is the correct choice. Choices B and C are inconsistent with Part 97.

G1E03 [97.221] What is required to conduct communications with a digital station operating under automatic control outside the automatic control band segments?
   A. The station initiating the contact must be under local or remote control
   B. The interrogating transmission must be made by another automatically controlled station
   C. No third-party traffic may be transmitted
   D. The control operator of the interrogating station must hold an Extra Class license

In this instance, Part 97 requires that the "station is responding to interrogation by a station under local or remote control." This rule makes **Answer A** the right choice.

G1E04 [97.13(b), 97.303, 97.311(b)] Which of the following conditions require a licensed Amateur Radio operator to take specific steps to avoid harmful interference to other users or facilities?
   A. When operating within one mile of an FCC Monitoring Station
   B. When using a band where the Amateur Service is secondary
   C. When a station is transmitting spread spectrum emissions
   D. All of these choices are correct

The Part 97 regulations explicitly call out each of the conditions specified in Answers A, B, and C for when to avoid causing harmful interference. Therefore, the best choice is **Answer D**.

G1E05 [97.115(a)(2),97.117] What types of messages for a third party in another country may be transmitted by an amateur station?
   A. Any message, as long as the amateur operator is not paid
   B. Only messages for other licensed amateurs
   C. Only messages relating to Amateur Radio or remarks of a personal character, or messages relating to emergencies or disaster relief
   D. Any messages, as long as the text of the message is recorded in the station log

Under Part 97, transmissions "to a different country, where permitted, shall be limited to communications incidental to the purposes of the amateur service and to remarks of a personal character." The correct choice is **Answer C** because it complies with the regulations. Answer B is incorrect because third-party communications are generally not for other licensed amateurs. Answers A and D are incorrect because "any message" might not fall within the restrictions given in Part 97.

G1E06 [97.301, ITU Radio Regulations] The frequency allocations of which ITU region apply to radio amateurs operating in North and South America?
  A. Region 4
  B. Region 3
  C. Region 2
  D. Region 1

If we look at Figure 1.1, we can see that the Americas are in Region 2. This figure makes **Answer C** the correct choice. Note: there is no Region 4, so do not be confused with that distraction answer.

G1E07 [97.111] In what part of the 13-centimeter band may an amateur station communicate with non-licensed Wi-Fi stations?
  A. Anywhere in the band
  B. Channels 1 through 4
  C. Channels 42 through 45
  D. No part

While parts of the Amateur Service bandshare with Wi-Fi communications, the general principle in Part 97 is that an amateur station exchanges two-way transmissions "necessary to exchange messages with other stations in the amateur service." Consumer digital Wi-Fi data communications are unlicensed and do not fall under the Amateur Service framework, so their users are not to communicate with stations in the Amateur Service. This Amateur Service restriction makes **Answer D** the correct choice.

G1E08 [97.313(j)] What is the maximum PEP output allowed for spread spectrum transmissions?
  A. 100 milliwatts
  B. 10 watts
  C. 100 watts
  D. 1500 watts

While we cannot use Wi-Fi equipment and protocols in their "factory mode" for amateur transmissions, we can re-purpose the equipment for other uses such as the Amateur Radio Emergency Data Network (AREDN) service. The Wi-Fi radios use a form of SS communications. Part 97 restricts SS stations such that "No station may transmit with a transmitter output exceeding 10 W PEP when the station is transmitting a SS emission type." This restriction makes **Answer B** the correct choice.

G1E09 [97.115] Under what circumstances are messages that are sent via digital modes exempt from Part 97 third-party rules that apply to other modes of communication?
  A. Under no circumstances
  B. When messages are encrypted
  C. When messages are not encrypted
  D. When under automatic control

When you inspect Part 97, you will find that there is no exemption made to the third-party rules for digital communications methods. The lack of a digital-mode exemption makes **Answer A** the correct choice.

G1E10 [97.101] Why should an amateur operator normally avoid transmitting on 14.100, 18.110, 21.150, 24. 930 and 28.200 MHz?
  A. A system of propagation beacon stations operates on those frequencies
  B. A system of automatic digital stations operates on those frequencies
  C. These frequencies are set aside for emergency operations
  D. These frequencies are set aside for bulletins from the FCC

Part 97 and good amateur practice hold that "[n]o amateur operator shall willfully or maliciously interfere with or cause interference to any radio communication or signal." The frequencies listed are common locations to find propagation beacon stations that help the whole Amateur Service community, so operators should avoid these frequencies, as **Answer A** states.

G1E11 [97.221, 97.305] On what bands may automatically controlled stations transmitting RTTY or data emissions communicate with other automatically controlled digital stations?
  A. On any band segment where digital operation is permitted
  B. Anywhere in the non-phone segments of the 10-meter or shorter wavelength bands
  C. Only in the non-phone Extra Class segments of the bands
  D. Anywhere in the 6-meter or shorter wavelength bands, and in limited segments of some of the HF bands

Part 97 states that an amateur "station may be automatically controlled while transmitting a RTTY or data emission on the 6 m or shorter wavelength bands, and on the 28.120-28.189 MHz, 24.925-24.930 MHz, 21.090-21.100 MHz, 18.105-18.110 MHz, 14.0950-14.0995 MHz, 14.1005-14.112 MHz, 10.140-10.150 MHz, 7.100-7.105 MHz, or 3.585-3.600 MHz segments." **Answer D** is correct because it complies with this rule.

Chapter 2

G2 — OPERATING PROCEDURES

2.1 Introduction

The General Class license will give you greater access to the Low Frequency (LF), Medium Frequency (MF) and High Frequency (HF) amateur bands than you have with the Technician license. You will also have a wider array of potential operating modes with those privileges. This subelement has many questions dealing with both the Federal Communications Commission (FCC) regulations from Part 97 and the mechanics of operating on the bands. These operational questions can teach you about correct procedures commonly found in the amateur community so that you will mesh well with other operators. The *Operating Procedures* subelement has the following question groups:

A. Phone Operations
B. On the Bands
C. CW Operations
D. HF Operations
E. Digital Operations

Subelement 2 will generate five questions on the General Class examination.

2.2 Radio Engineering Concepts

Modulation Bandwidth Review The General Class license exam bases several questions on the emission bandwidth for the different operating modes. Part 97 defines *bandwidth* as the "width of a frequency band outside of which the mean power of the transmitted signal is attenuated at least 26 dB below the mean power of the transmitted signal within the band." Engineers use this definition because the signals do not have well-defined edges in the frequency band. With this definition, the signal strength outside the band is at least 400 times weaker than the average power in the band. In the Technician Class study guide, we saw that the frequency space needed to transmit a signal depends on the underlying signal's characteristics and the modulation mode we use when transmitting. When using Single Sideband (SSB), you need to remember that Continuous Wave (CW) requires approximately 150 Hz, while phone requires approximately 2 kHz to 3 kHz. The bandwidth scales with the transmitted digital modulation symbol rate for digital bandwidth estimates.

Amplitude Modulation Review   As we saw in the Technician Class study guide, Amplitude Modulation (AM) uses the message signal to modify the carrier's amplitude. AM comes in two major classes:

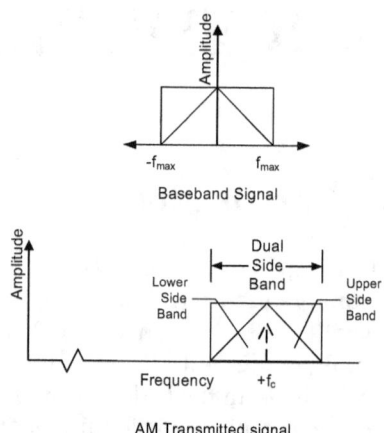

Single Sideband — uses one copy of the message signal's frequency domain content in the transmission

Dual Sideband — uses both copies of the message signal's frequency domain content in the transmission

SSB comes in two modes: Upper Side Band (USB) and Lower Side Band (LSB). Dual Sideband (DSB) also comes in two modes: Dual Sideband - Residual Carrier (DSB-RC) where the transmitter sends an unmodulated copy of the carrier to aid in reception, and Dual Sideband - Suppressed Carrier (DSB-SC) where the

Figure 2.1: Amplitude modulation spectrum for DSB and SSB relative to the carrier f_c.

transmitter does not send a copy of the carrier. DSB-RC is the format used in commercial broadcast AM transmissions. Figure 2.1 shows the difference between SSB and DSB in the frequency domain. The *baseband* block is the message signal as seen by a spectrum analyzer. When the message signal amplitude modulates the carrier, a copy of the baseband spectrum appears in frequency space at the carrier location, f_c. DSB transmits the full copy of the spectrum. LSB transmits one half of the spectrum below the carrier, while USB transmits one half of the spectrum above the carrier. Neither SSB mode transmits a copy of the carrier. Both sidebands contain all the necessary information to recover the message signal in the receiver. SSB uses less transmission bandwidth than DSB at the expense of a bit more complicated electronics.

Table 2.1: Selected Q Signals for the General Class Exam.

Signal	Meaning
QRL	Are you busy?
QRM	Do you have interference? [from other stations]
QRN	Are you troubled by static?
QRO	Shall I increase power?
QRP	Shall I decrease power?
QRS	Shall I send slower?
QRU	Have you anything for me?
QRV	Are you ready?
QRZ	Who is calling me?
QSB	Are my signals fading?
QSK	Can you work break in?
QSL	Can you acknowledge receipt?
QSO	Can you communicate with?
QSY	Shall I change to transmission on another frequency?
QSZ	Shall I send each word or group more than once?
QTH	What is your position?

Q Signals   In this chapter, we have more questions on the "Q signals." Table 2.1 lists even more of the common Q signals you may wish to memorize to help with your operating techniques on the air during your QSO with another station. You can find an extensive list at `https://en.wikipedia.org/wiki/Q_code`.

Procedural Signals and Abbreviations   With CW and digital modes, the operators frequently use standard *procedural signals* ("prosigns") and abbreviations to speed up transmission. Table 2.2 lists common prosigns you will need for the license exam plus some useful abbreviations for operating on the bands.

RST Reports   Operators often give reception reports to the transmitting station so that the sender can know how well the operator is receiving their signal. Generally, amateurs use the Readability-Signal Strength-Tone (RST) method. The RST classification is

Readability — a rating from 1 to 5, with 1 meaning the transmission was unreadable and 5 meaning perfectly readable

Signal Strength — a rating from 1 through 9, with 1 meaning faint signals that are barely detectable and 9 meaning extremely strong signals

Tone — a rating from 1 to 9, with one meaning the tone is very rough with modulation on the signal, and 9 means perfect tone with no residual modulation

The RST report was designed for CW transmissions. However, you will also encounter it with phone and digital modes. In the latter cases, the Tone measure-

Table 2.2: Selected Prosigns and Abbreviations.

Prosign	Meaning	Abbrev.	Meaning
AR	End of current message	AGN	Again
BK	Receiving station to answer now	DX	Distance
CL	Shutting down station completely	LID	Poor operator
CQ	Calling any Amateur Service station	HI	Laugh
K	Go ahead	OM	Old Man
KN	Specified station go ahead	TU	Thank You
R	Received all	WX	Weather
SK	End of contact	73	Best Regards

ment is irrelevant, so operators often ignore it. As a shorthand, many operators send a "599" to indicate they received it all without giving the fine-detailed information. Operators can also add a suffix to the report with a "C" for signal chirp, "K" for clicks, or "X" for very stable frequency.

Digital Operating Modes   The availability of computers with sound cards and the internet to distribute software has increased available digital transmission modes. These modes use the computer's sound card to process the digital data's transmission and reception. The computer exchanges the data with the HF rig for transmission over the air. The software displays the transmission on a *waterfall display* that shows the individual transmissions across the frequency band. The waterfall is a running scroll in time to trace the ongoing transmissions. The individual messages are shown horizontally across the frequency dimension. The vertical dimension illustrates activity in time. Figure 2.2 shows a PSK31 waterfall display with multiple transmissions. Similarly, Figure 2.3 shows an FT8 waterfall display with multiple transmissions.

Some of the available modes that the various software packages can support are

CW — many of the software packages support Morse code, which is the original digital mode

FT8, JT4, JT9, JT65, WSPR — a family of low data rate transmission methods designed for reliable text exchanges under very weak signal/low-power conditions, including moonbounce, meteor scatter, and propagation probing in addition to supporting normal exchanges

Hellschreiber — a scanning mode, like with a dot matrix printer, to print the text as a picture

PSK31 — digital text transmissions that modulates a Phase Shift Keying (PSK) signal; 31, 63, and 125 bps are typical data rates

MFSK — digital transmission for text and images by Multiple Frequency Shift Keying (MFSK) carriers with included error correction coding that permits operation in bad conditions; there are several variants of this technique,

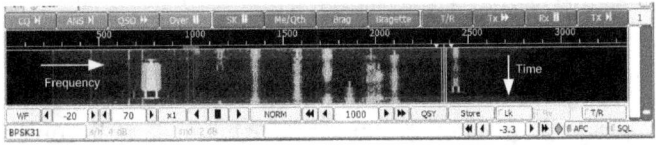

Figure 2.2: Waterfall display from digital transmission software showing PSK31 transmissions.

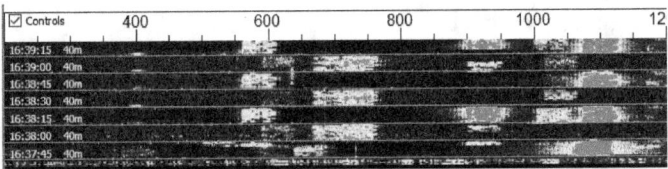

Figure 2.3: Waterfall display from digital transmission software showing FT8 transmissions. Note the time marks on the left-hand edge of the display.

such as Olivia and Contestia, that have better operating characteristics (less need for exact tuning, etc.)

PACTOR — a packet radio technique using both error control techniques and automatic retransmission requests; requires an external modem

RTTY — Radio Teletype for digital text transmission

WINMOR — similar to PACTOR, but can run over a computer sound card

All the digital modes, except FT8, JT4, JT9, and JT65, are designed to transmit regular text-based conversations between the operators. These conversations can last as long as the operators wish. The FT/JT family is more like a contest with minimal, essential information exchanged. Figure 2.4 shows an example FT8 contact where WA5DJJ at grid location DM62 is calling CQ. NM4SH at grid location FM17 responds. WA5DJJ acknowledges with a signal report of -03, while NM4SH replies with a signal report of -04. The operators acknowledge the information exchange and send their 73s to complete the QSO. *Note:* the signal reports here reference noise levels and not RST reports.

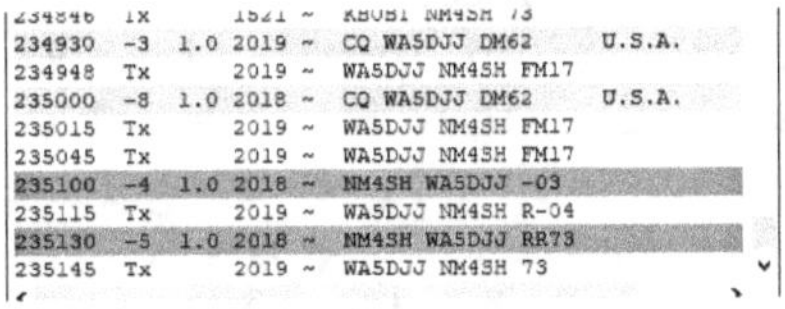

```
234846  1X      1621 ~  KBUB1 NM4SH 73
234930  -3  1.0 2019 ~  CQ WA5DJJ DM62      U.S.A.
234948  Tx      2019 ~  WA5DJJ NM4SH FM17
235000  -8  1.0 2018 ~  CQ WA5DJJ DM62      U.S.A.
235015  Tx      2019 ~  WA5DJJ NM4SH FM17
235045  Tx      2019 ~  WA5DJJ NM4SH FM17
235100  -4  1.0 2018 ~  NM4SH WA5DJJ -03
235115  Tx      2019 ~  WA5DJJ NM4SH R-04
235130  -5  1.0 2018 ~  NM4SH WA5DJJ RR73
235145  Tx      2019 ~  WA5DJJ NM4SH 73
```

Figure 2.4: Typical FTY8 contact between WA5DJJ and NM4SH.

2.3   G2A - Phone Operations

2.3.1   Overview

The *Phone Operations* question group in Subelement G2 tests you on operating standards used in phone transmission. The *Phone Operations* group covers topics such as

- Phone operating procedures
- USB/LSB conventions
- Breaking into a contact
- VOX operation

The test producer will select one of the 12 questions in this group for your exam.

2.3.2   Questions

G2A01 Which sideband is most commonly used for voice communications on frequencies of 14 MHz or higher?
  A. Upper sideband
  B. Lower sideband
  C. Vestigial sideband
  D. Double sideband

Amateur operational practice for voice communications that has grown up over the years, not Part 97 rules, is the basis for this question. If you have not participated in HF operations, it might not make much sense until you get used to the procedure. The accepted practice is that below 14 MHz, or bands longer than 20 m, use LSB. At or above 14 MHz, or at the 20-m band and shorter wavelengths, use USB. Later, when we discuss the 60-m band, we will see a regulatory difference to this general rule. Since we are on or above 20 m, we choose USB, and the correct choice is **Answer A**. Vestigial Side Band (VSB) is a mode designed for TV transmission and not phone transmission. DSB is standard AM, and operators do not commonly use it on these bands.

G2A02 Which of the following modes is most commonly used for voice communications on the 160-meter, 75-meter, and 40-meter bands?
  A. Upper sideband
  B. Lower sideband
  C. Vestigial sideband
  D. Double sideband

Using the operating practice from the previous question, we can spot that the right answer is LSB, as in **Answer B**, on these bands. Answer A is for higher frequencies, so it is incorrect. The other two options are to distract you.

G2A03 Which of the following is most commonly used for SSB voice communications in the VHF and UHF bands?
A. Upper sideband
B. Lower sideband
C. Vestigial sideband
D. Double sideband

Because the Very High Frequency (VHF) and Ultra High Frequency (UHF) bands are above 20 m in frequency, we can see that the right choice is USB, as in **Answer A**. Answer B violates the operational protocol in these bands, so it is incorrect. Answers C and D are still there to distract you.

G2A04 Which mode is most commonly used for voice communications on the 17-meter and 12-meter bands?
A. Upper sideband
B. Lower sideband
C. Vestigial sideband
D. Double sideband

Because 17 m and 12 m are higher in frequency than the 20 m band, the correct choice is USB, which makes **Answer A** the right choice.

G2A05 Which mode of voice communication is most commonly used on the HF amateur bands?
A. Frequency modulation
B. Double sideband
C. Single sideband
D. Phase modulation

Since we have been stressing SSB with the past few questions, you may suspect that the correct answer is SSB, as in **Answer C**, and you are correct. But why? In the HF bands, the operating frequency space is limited, so we wish to have the smallest "footprint" on the band. For phone operation, Frequency Modulation (FM) and Phase Modulation (PM) will have the widest bandwidth at around 15 kHz, AM is the next widest at around 6 kHz, and SSB is the narrowest choice at around 3 kHz. Therefore, SSB is the best choice operationally.

G2A06 Which of the following is an advantage when using single sideband, as compared to other analog voice modes on the HF amateur bands?
A. Very high fidelity voice modulation
B. Less subject to interference from atmospheric static crashes
C. Ease of tuning on receive and immunity to impulse noise
D. Less bandwidth used and greater power efficiency

FM and PM are superior to SSB in static protection and voice transmission quality, so Answers A and B are incorrect. Many modulations are similar in tuning ease with modern receivers, so Answer C is not a good choice. However, SSB is superior in bandwidth and power efficiency, so **Answer D** is correct.

G2A07 Which of the following statements is true of the single sideband voice mode?
A. Only one sideband and the carrier are transmitted; the other sideband is suppressed
B. Only one sideband is transmitted; the other sideband and carrier are suppressed
C. SSB is the only voice mode that is authorized on the 20-meter, 15-meter, and 10-meter amateur bands
D. SSB is the only voice mode that is authorized on the 160-meter, 75-meter and 40-meter amateur bands

From Figure 2.1, we can see that SSB has the designated sideband transmitted without a carrier signal, as in **Answer B**. Answer A is incorrect because it states that the carrier is transmitted. If you look at Table 1.1 in Chapter 1, you will see other modes than SSB are permitted, so Answers C and D are also incorrect.

G2A08 What is the recommended way to break in to a phone contact?
A. Say "QRZ" several times followed by your call sign
B. Say your call sign once
C. Say "Breaker Breaker"
D. Say "CQ" followed by the call sign of either station

When you send "QRZ," you request the station's call sign, so Answer A is not a good choice. Saying your call sign once, as in **Answer B**, is the correct operating procedure. Answer C is bad amateur operational practice, so it is incorrect. When you send "CQ," you are looking for another station, not trying to enter an existing conversation, so Answer D is not a good choice.

G2A09 Why do most amateur stations use lower sideband on the 160-meter, 75-meter and 40-meter bands?
A. Lower sideband is more efficient than upper sideband at these frequencies
B. Lower sideband is the only sideband legal on these frequency bands
C. Because it is fully compatible with an AM detector
D. It is good amateur practice

Answers A and C are technobabble because the sidebands have equal efficiency, and SSB generally cannot be detected with a dual sideband AM receiver. Answer B is incorrect because Part 97 regulations do not specify the sideband mode. The current amateur practice of **Answer D** is the correct choice.

G2A10 Which of the following statements is true of voice VOX operation versus PTT operation?
  A. The received signal is more natural sounding
  B. It allows "hands free" operation
  C. It occupies less bandwidth
  D. It provides more power output

A Voice Operated Switch (VOX) keys the transmitter like the Push to Talk (PTT) switch does on the microphone. The only effect this relay can have is to allow you to have a "hands-free" operation. It cannot process or change the signal in any way. Therefore, **Answer B** is the best choice based on how radios operate.

G2A11 Generally, who should respond to a station in the contiguous 48 states who calls "CQ DX"?
  A. Any caller is welcome to respond
  B. Only stations in Germany
  C. Any stations outside the lower 48 states
  D. Only contest stations

The "CQ" means you are looking for a station, while the "DX" implies a distant (not one in the caller's country) station. Therefore "CQ DX" is calling for a distant station, as in **Answer C**. Note: for the "lower 48," amateurs consider Alaska, Hawaii, and off-shore US territories to be "DX" stations. Answer A would be correct if we are just sending "CQ." Germany uses DA through DR prefixes.

G2A12 What control is typically adjusted for proper ALC setting on an amateur single sideband transceiver?
  A. The RF clipping level
  B. Transmit audio or microphone gain
  C. Antenna inductance or capacitance
  D. Attenuator level

The Automatic Level Control (ALC) circuit adjusts the audio or microphone input level, so **Answer B** is the right choice. The other choices are distractions.

2.4   G2B - On the Bands

2.4.1   Overview

The *On the Bands* question group in Subelement G2 quizzes you on good amateur radio practice when operating your station. The *On the Bands* group covers topics such as
  • Operating courtesy

- Band plans
- Emergencies, including drills and emergency communications

The test producer will select one of the 11 questions in this group for your exam.

2.4.2   Questions

G2B01 Which of the following is true concerning access to frequencies?
- A. Nets always have priority
- B. QSOs in progress always have priority
- C. Except during emergencies, no amateur station has priority access to any frequency
- D. Contest operations must always yield to non-contest use of frequencies

Part 97 states (a) "No frequency will be assigned for the exclusive use of any station." and (b) "At all times and on all frequencies, each control operator must give priority to stations providing emergency communications, except to stations transmitting communications for training drills and tests in RACES." **Answer C** is the correct choice, and the others are inconsistent with Part 97.

G2B02 What is the first thing you should do if you are communicating with another amateur station and hear a station in distress break in?
- A. Continue your communication because you were on the frequency first
- B. Acknowledge the station in distress and determine what assistance may be needed
- C. Change to a different frequency
- D. Immediately cease all transmissions

If we are to fulfill the goal of the amateur service to be a national resource in times of distress, and if we exercise a bit of common sense, we conclude that **Answer B** is the best choice. Answers A and C will not help the station in distress. Answer D may sound correct, but then how could you help?

G2B03 What is good amateur practice if propagation changes during a contact and you notice interference from other stations on the frequency?
- A. Tell the interfering stations to change frequency
- B. Report the interference to your local Amateur Auxiliary Coordinator
- C. Attempt to resolve the interference problem with the other stations in a mutually acceptable manner
- D. Increase power to overcome interference

Best amateur operating practice says that the interfering operators should mutually resolve the situation, as in **Answer C**. One possible action to take would be sending a QSY and changing frequencies. Answer A violates the rule that no one has exclusive access to a frequency. Answer B will not eliminate the interference. Answer D is bad operating practice, so it is incorrect. *Note:* the

Amateur Auxiliary function is now the Volunteer Monitoring Program (VMP).

G2B04 When selecting a CW transmitting frequency, what minimum separation should be used to minimize interference to stations on adjacent frequencies?
A. 5 to 50 Hz
B. 150 to 500 Hz
C. 1 to 3 kHz
D. 3 to 6 kHz

Here, you need to know that CW transmissions have a bandwidth around 150 Hz, so **Answer B** gives the minimum user frequency separation. We will see more about signal bandwidths in Chapter 8 and Table 8.2. Answer A has the signals overlap, and we will still have interference. Answers C and D are unnecessarily large separations and are not as good choices as Answer B.

G2B05 When selecting an SSB transmitting frequency, what minimum separation should be used to minimize interference to stations on adjacent frequencies?
A. 5 to 50 Hz
B. 150 to 500 Hz
C. Approximately 3 kHz
D. Approximately 6 kHz

This question is the same type as the previous one. With this one, you need to know that SSB transmissions have bandwidths around 3 kHz (Table 8.2). Therefore, **Answer C** gives the minimum separation between the users, so it is the best choice of the ones given. Answer B would have the users' signals overlap because it is for CW and we would still have interference. Answers D is for DSB and not for SSB. Answer A is to distract you.

G2B06 What is a practical way to avoid harmful interference on an apparently clear frequency before calling CQ on CW or phone?
A. Send "QRL?" on CW, followed by your call sign; or, if using phone, ask if the frequency is in use, followed by your call sign
B. Listen for 2 minutes before calling CQ
C. Send the letter "V" in Morse code several times and listen for a response, or say "test" several times and listen for a response
D. Send "QSY" on CW or if using phone, announce "the frequency is in use", then give your call and listen for a response

Each of the actions listed in Answers B, C, and D is individually a bad operating practice or a silly distraction, so they are incorrect choices. **Answer A** gives the correct procedures for CW and phone, so this is the one to choose.

G2B07 Which of the following complies with good amateur practice when choosing a frequency on which to initiate a call?
  A. Check to see if the channel is assigned to another station
  B. Identify your station by transmitting your call sign at least 3 times
  C. Follow the voluntary band plan for the operating mode you intend to use
  D. All of these choices are correct

Answer A is incorrect since there are no formal assignments in the amateur service. Answer B is wrong because it does not help choose an operating frequency. Since Answers A and B are incorrect, Answer D is also incorrect. Following the voluntary band plan, as mentioned in **Answer C**, is the procedure for good amateur practice. You can find a band plan listing from the ARRL at http://www.arrl.org/considerate-operator.

G2B08 What is the voluntary band plan restriction for U.S. stations transmitting within the 48 contiguous states in the 50.1 to 50.125 MHz band segment?
  A. Only contacts with stations not within the 48 contiguous states
  B. Only contacts with other stations within the 48 contiguous states
  C. Only digital contacts
  D. Only SSTV contacts

The 6-m band plan shows that the 50.1 MHz to 50.125 MHz sub-band is the "DX Window" for the band. The band plan reserves this region for U.S. operators within the 48 contiguous states to contact stations outside the 48 contiguous states. This definition makes **Answer A** the correct choice. Be careful because Answer B is the opposite of the DX sense. The other options are to distract you.

G2B09 [97.407(a)] Who may be the control operator of an amateur station transmitting in RACES to assist relief operations during a disaster?
  A. Only a person holding an FCC issued amateur operator license
  B. Only a RACES net control operator
  C. A person holding an FCC issued amateur operator license or an appropriate government official
  D. Any control operator when normal communication systems are operational

The Part 97 rules for Radio Amateur Civil Emergency Service (RACES) state that "[n]o person may be the control operator of a RACES station, or may be the control operator of an amateur station transmitting in RACES unless that person holds a FCC-issued amateur operator license and is certified by a civil defense organization as enrolled in that organization." **Answer A** meets the regulations. Be careful with Answer C because its phrasing does not match the rules. Answers B and D are distractions.

G2B10 [97.407(b)] When is an amateur station allowed to use any means at its disposal to assist another station in distress?
A. Only when transmitting in RACES
B. At any time when transmitting in an organized net
C. At any time during an actual emergency
D. Only on authorized HF frequencies

Looking again to Part 97, we see that "[n]o provision of these rules prevents the use by a station ... of any means of radiocommunications at its disposal to assist a station in distress." This rule matches **Answer C**, making it correct.

G2B11 [97.405] What frequency should be used to send a distress call?
A. Whichever frequency has the best chance of communicating the distress message
B. Only frequencies authorized for RACES or ARES stations
C. Only frequencies that are within your operating privileges
D. Only frequencies used by police, fire or emergency medical services

As we just saw for emergencies, Part 97 lets you can use whatever frequency allows you to communicate when in distress. This rule makes **Answer A** the right choice for this question. The other choices would place restrictions on you that are not in Part 97, so they represent distractions.

2.5 G2C - CW Operations

2.5.1 Overview

The *CW Operations* question group in Subelement G2 covers operating practice for CW exchanges. Digital exchanges also use some of these practices. The **CW Operations** group covers topics such as
- CW operating procedures and procedural signals
- Q signals and common abbreviations
- Full break-in telegraphy
The test producer will select one of the 11 questions in this group for your exam.

2.5.2 Questions

G2C01 Which of the following describes full break-in telegraphy (QSK)?
A. Breaking stations send the Morse code prosign "BK"
B. Automatic keyers, instead of hand keys, are used to send Morse code
C. An operator must activate a manual send/receive switch before and after every transmission
D. Transmitting stations can receive between code characters and elements

As you gain more experience with HF rigs, you will recognize that receiving CW between code characters is the definition for QSK, as in **Answer D**. Answer C would prolong operations and achieves the opposite effect of QSK. Answers A and B might sound reasonable, but they do not fit the definition.

G2C02 What should you do if a CW station sends "QRS"?
- A. Send slower
- B. Change frequency
- C. Increase your power
- D. Repeat everything twice

"QRS" is the Q signal for CW transmission speed. In this case, it means the operator should send more slowly, so **Answer A** is the correct choice. Answer B corresponds to "QSY." Answer C corresponds to "QRO" (the opposite of "QRP"). Answer D is "QSZ."

G2C03 What does it mean when a CW operator sends "KN" at the end of a transmission?
- A. Listening for novice stations
- B. Operating full break-in
- C. Listening only for a specific station or stations
- D. Closing station now

"KN" is a procedural signal inviting a specific station or stations to transmit. This definition makes **Answer C** the correct choice. Answer D is "CL." As we saw above, operating full break-in is "QSK." Answer A is to make you smile.

G2C04 What does the Q signal "QRL?" mean?
- A. "Will you keep the frequency clear?"
- B. "Are you operating full break-in?" or "Can you operate full break-in?"
- C. "Are you listening only for a specific station?"
- D. "Are you busy?", or "Is this frequency in use?"

This new procedural sign, "QRL," is asking if the frequency is in use or if the operator is busy, as in **Answer D**. The other choices are to distract you.

G2C05 What is the best speed to use when answering a CQ in Morse code?
- A. The fastest speed at which you are comfortable copying, but no slower than the CQ
- B. The fastest speed at which you are comfortable copying, but no faster than the CQ
- C. At the standard calling speed of 10 wpm
- D. At the standard calling speed of 5 wpm

This question seeks a common-sense solution to match the abilities of both

operators. We assume that you can copy at your sending speed, so this is the
top speed. Similarly, you do not want to send faster than the CQ speed. Putting
these two together gives **Answer B** as the correct choice. Answer A might be
good for you, but not the sender. The other choices are distractions.

G2C06 What does the term "zero beat" mean in CW operation?
   A. Matching the speed of the transmitting station
   B. Operating split to avoid interference on frequency
   C. Sending without error
   D. Matching your transmit frequency to the frequency of a received signal

The term "zero beat" means the receiver is matching the frequency of the transmitting station, as in **Answer D**. This is done because, if the frequencies are not
matched, there will be a beating of the signals that can be heard as a tone. The
other choices do not have anything to do with this term.

G2C07 When sending CW, what does a "C" mean when added to the RST
report?
   A. Chirpy or unstable signal
   B. Report was read from an S meter rather than estimated
   C. 100 percent copy
   D. Key clicks

In addition to readability, signal strength, and tone, there are times when there
are other common problems that the operator needs to note in the RST report.
The notation "C" means signal chirp is present. This notation makes **Answer A**
the correct choice. Answer D corresponds to a "K" in the RST report. Answer B
is friendly but irrelevant. Operators often report copy results, but it is not an
official part of the RST report.

G2C08 What prosign is sent to indicate the end of a formal message when
using CW?
   A. SK
   B. BK
   C. AR
   D. KN

The correct *prosign* (procedural sign) for the end of a formal message is "AR,"
so **Answer C** is the correct choice. "SK" means the end of QSO. "BK" means
BacK to you. We saw above that "KN" is for specific stations to transmit.

G2C09 What does the Q signal "QSL" mean?
- A. Send slower
- B. We have already confirmed by card
- C. I acknowledge receipt
- D. We have worked before

The Q signal "QSL" means "I copy" or "I acknowledge receipt," so **Answer C** is the correct choice. We saw above that "QRS" means send slower. Be careful with Answer B because you may soon be collecting QSL cards, but that is not implied by "QSL" during transmission. Answer D is nice, but not a Q signal.

G2C10 What does the Q signal "QRN" mean?
- A. Send more slowly
- B. Stop sending
- C. Zero beat my signal
- D. I am troubled by static

The Q signal "QRN" means that you are receiving static, as in **Answer D**. Operators use "QRS" to request slower transmission.

G2C11 What does the Q signal "QRV" mean?
- A. You are sending too fast
- B. There is interference on the frequency
- C. I am quitting for the day
- D. I am ready to receive messages

The Q signal "QRV" means that you are ready for messages, so **Answer D** is the correct choice. Operators use "QRS" for Answer A. Answer B is "QRM." Answer C is the prosign "CL" and not a Q signal.

2.6   G2D - HF Operations

2.6.1   Overview

The *HF Operations* question group in Subelement G2 introduces you to basic electrical principles found in radio circuits. The *HF Operations* group covers topics such as
- Volunteer Monitoring Program
- HF operations

The test producer will select one of the 11 questions in this group for your exam.

2.6.2 Questions

G2D01 What is the Volunteer Monitoring Program?
  A. Amateur volunteers who are formally enlisted to monitor the airwaves for rules violations
  B. Amateur volunteers who conduct amateur licensing examinations
  C. Amateur volunteers who conduct frequency coordination for amateur VHF repeaters
  D. Amateur volunteers who use their station equipment to help civil defense organizations in times of emergency

The VMP is the activity where amateurs assist the FCC as in **Answer A**. Answer B is for the Volunteer Examiners (VEs), Answer C is a frequency coordinator, and Answer D is RACES or Amateur Radio Emergency Service (ARES).

G2D02 Which of the following are objectives of the Volunteer Monitoring Program?
  A. To conduct efficient and orderly amateur licensing examinations
  B. To encourage amateur radio operators to self-regulate and comply with the rules
  C. To coordinate repeaters for efficient and orderly spectrum usage
  D. To provide emergency and public safety communications

Using the previous question, we should be able to eliminate Answers A, C, and D from consideration. The correct choice is **Answer B** because it describes the self-regulation of the Amateur Radio Service by the amateur radio community.

G2D03 What skills learned during hidden transmitter hunts are of help to the Volunteer Monitoring Program?
  A. Identification of out-of-band operation
  B. Direction finding used to locate stations violating FCC Rules
  C. Identification of different call signs
  D. Hunters have an opportunity to transmit on non-amateur frequencies

Hidden transmitter hunts, also known as fox hunts, have nothing to do with animals. Instead, it is a direction-finding skill, so the correct choice is **Answer B**. Answer D is to make you smile, while Answers A and C are irrelevant.

G2D04 Which of the following describes an azimuthal projection map?
  A. A map that shows accurate land masses
  B. A map that shows true bearings and distances from a particular location
  C. A map that shows the angle at which an amateur satellite crosses the equator
  D. A map that shows the number of degrees longitude that an amateur satellite appears to move westward at the equator with each orbit

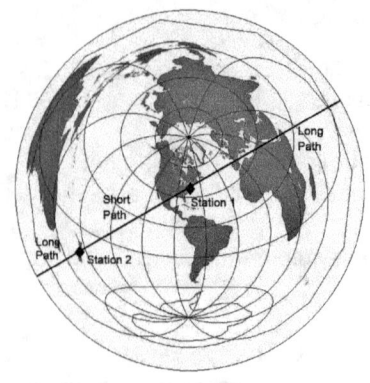

The azimuthal map is a world map centered on a specific location, as in **Answer B**. This map helps determine which way to point a directional antenna to reach a DX location. All the other answers are there as distractions. Figure 2.5 illustrates an azimuthal map centered in southeastern Virginia with two stations marked by diamonds. The straight line indicates the short and long propagation paths between them.

Figure 2.5: Example of an azimuthal projection world map.

G2D05 Which of the following is a good way to indicate on a clear frequency in the HF phone bands that you are looking for a contact with any station?
  A. Sign your call sign once, followed by the words "listening for a call" – if no answer, change frequency and repeat
  B. Say "QTC" followed by "this is" and your call sign – if no answer, change frequency and repeat
  C. Repeat "CQ" a few times, followed by "this is," then your call sign a few times, then pause to listen, repeat as necessary
  D. Transmit an unmodulated carried for approximately 10 seconds, followed by "this is" and your call sign, and pause to listen – repeat as necessary

The key to correctly answering this question is to start with the "CQ" and then follow with your call sign, as in **Answer C**. The other choices do not represent good amateur practice, so they are incorrect.

G2D06 How is a directional antenna pointed when making a "long-path" contact with another station?
  A. Toward the rising Sun
  B. Along the grayline
  C. 180 degrees from its short-path heading
  D. Toward the north

The radio path between the two stations can directly follow the shortest arc between the stations. It can also travel "the long way around" and take the "long path" arc around the backside of the Earth as Figure 2.5 illustrates. By looking at the shape and geometry of a sphere, we can see that short paths and long paths point 180 degrees apart, so the correct choice is **Answer C**. The gray line is the line following the sunrise or sunset lines, so it may not even be pointing along the propagation path.

G2D07 Which of the following are examples of the NATO Phonetic Alphabet?
  A. Able, Baker, Charlie, Dog
  B. Adam, Boy, Charles, David
  C. America, Boston, Canada, Denmark
  D. Alpha, Bravo, Charlie, Delta

There are several standard phonetic alphabets, but none start with "Adam" or "America," so we eliminate Answers B and C. The NATO phonetic alphabet, also known as the International Radiotelephony Spelling Alphabet, starts with "Alpha," so **Answer D** is the correct choice. The US military used the "Able-Baker" alphabet until the 1950s, so it is incorrect now.

Date	Freq.	Mode	Power	Time On	Station	Sent	Rcvd	Time Off	Qth/Grid	Name	Notes

Figure 2.6: Example station log sheet.

G2D08 What is a reason why many amateurs keep a station log?
  A. The ITU requires a log of all international contacts
  B. The ITU requires a log of all international third-party traffic
  C. The log provides evidence of operation needed to renew a license without retest
  D. To help with a reply if the FCC requests information

If the FCC questions some aspect of your station's operation, it is helpful to have a written record, or operational log, for reference. Figure 2.6 shows an example of a paper log; there are computer-based logs too. This log makes **Answer D** the best choice. The International Telecommunication Union (ITU) does not have jurisdiction over your station, so Answers A and B are incorrect. The station log is not part of the licensing procedure, so Answer C is incorrect.

G2D09 Which of the following is required when participating in a contest on HF frequencies?
  A. Submit a log to the contest sponsor
  B. Send a QSL card to the stations worked, or QSL via Logbook of The World
  C. Identify your station per normal FCC regulations
  D. All these choices are correct

Every contest has different rules you must be familiar with when participating. The one thing they all have in common is that you need to correctly identify your station, which makes **Answer C** the correct choice. A QSL is a nicety but not a requirement. Submitting a log is typical to obtain an official contest score, but one does need to do this to participate. Answer D is to distract you.

G2D10 What is QRP operation?
   A. Remote piloted model control
   B. Low power transmit operation
   C. Transmission using Quick Response Protocol
   D. Traffic relay procedure net operation

The correct choice is **Answer B** because QRP is used to designate low-power operations. The other options are silly distractions.

G2D11 Which of the following is typical of the lower HF frequencies during the summer?
   A. Poor propagation at any time of day
   B. World-wide propagation during the daylight hours
   C. Heavy distortion on signals due to photon absorption
   D. High levels of atmospheric noise or "static"

Low HF frequencies are good for picking up static from thunderstorm activity, which is common during the summer, so **Answer D** is the correct choice. While daytime propagation is not favorable, nighttime propagation still is possible, so Answers A and B are incorrect. Answer C is a silly distraction.

2.7   G2E - Digital Operations

2.7.1   Overview

The *Digital Operations* question group in Subelement G2 introduces you to basic electrical principles found in radio circuits. The *Digital Operations* group covers digital methods, including procedures, procedural signals, and common abbreviations. The test producer will select one of the 15 questions in this group for your exam.

2.7.2   Questions

G2E01 Which mode is normally used when sending RTTY signals via AFSK with an SSB transmitter?
   A. USB
   B. DSB
   C. CW
   D. LSB

This question covers an operational protocol that will make better sense when you have more experience. You need to know that the correct response is LSB, as in **Answer D**. Using USB can flip the sense of the Audio Frequency Shift Keying (AFSK) tones with Radio TeleType (RTTY) operations. The other choices are distractions.

G2E02 How can a PACTOR modem or controller be used to determine if the channel is in use by other PACTOR stations?
   A. Unplug the data connector temporarily and see if the channel-busy indication is turned off
   B. Put the modem or controller in a mode which allows monitoring communications without a connection
   C. Transmit UI packets several times and wait to see if there is a response from another PACTOR station
   D. Send the message: "Is this frequency in use?"

For most modes, the correct amateur practice is to listen first before transmitting. The PACTOR equivalent is to silently monitor the channel before attempting any transmissions, which makes **Answer B** the right choice. Be careful with Answer D because it sounds reasonable, but it is still incorrect for this mode.

G2E03 What symptoms may result from other signals interfering with a PACTOR or WINMOR transmission?
   A. Frequent retries or timeouts
   B. Long pauses in message transmission
   C. Failure to establish a connection between stations
   D. All of these choices are correct

Each of the problems listed in Answers A, B, and C is a possibility in this situation, so the best choice is **Answer D**.

G2E04 What segment of the 20-meter band is most often used for digital transmissions (avoiding the DX propagation beacons)?
   A. 14.000 - 14.050 MHz
   B. 14.070 - 14.112 MHz
   C. 14.150 - 14.225 MHz
   D. 14.275 - 14.350 MHz

This question also covers radio practices that you will be more familiar with as you gain operating experience. From the 20-m band plan, we find that the correct choice is **Answer B** in the 14.070 MHz to 14.112 MHz segment just above the lower edge of the band. Be sure to avoid any propagation beacons at 14.100 MHz!

G2E05 What is the standard sideband used to generate a JT65, JT9, or FT8 digital signal when using AFSK in any amateur band?
- A. LSB
- B. USB
- C. DSB
- D. SSB

These are popular digital modes to try. JT65, JT9, and FT8 use USB, as in **Answer B**. Notice these modes do not obey the USB/LSB convention for phone communications.

G2E06 What is the most common frequency shift for RTTY emissions in the amateur HF bands?
- A. 85 Hz
- B. 170 Hz
- C. 425 Hz
- D. 850 Hz

You need to remember that, by standard practice, the correct choice for RTTY is 170 Hz, as in **Answer B**.

G2E07 What segment of the 80-meter band is most commonly used for digital transmissions?
- A. 3570 - 3600 kHz
- B. 3500 - 3525 kHz
- C. 3700 - 3750 kHz
- D. 3775 - 3825 kHz

Here is another radio practice question. Generally, the data segment of the band is near the low-frequency end but not at the lowest frequency segment. From the 80-m band plan, we see that the correct choice is **Answer A**, or the 3570 kHz to 3600 kHz segment. The other Answers are to test if you are familiar with the band plan.

G2E08 In what segment of the 20-meter band are most PSK31 operations commonly found?
- A. At the bottom of the slow-scan TV segment, near 14.230 MHz
- B. At the top of the SSB phone segment, near 14.325 MHz
- C. In the middle of the CW segment, near 14.100 MHz
- D. Below the RTTY segment, near 14.070 MHz

PSK31 is a digital data transmission mode. A popular place to find PSK31 transmissions is right around 14.070 MHz, so **Answer D** is the right choice. The others are not consistent with amateur operating practice.

G2E09 How do you join a contact between two stations using the PACTOR protocol?
- A. Send broadcast packets containing your call sign while in MONITOR mode
- B. Transmit a steady carrier until the PACTOR protocol times out and disconnects
- C. Joining an existing contact is not possible, PACTOR connections are limited to two stations
- D. Send a NAK response continuously so that the sending station must stand by

PACTOR is not like phone exchanges because there is no way to join the conversation. PACTOR is a point-to-point exchange between only two stations, so **Answer C** is the choice matching operating practice.

G2E10 Which of the following is a way to establish contact with a digital messaging system gateway station?
- A. Send an email to the system control operator
- B. Send QRL in Morse code
- C. Respond when the station broadcasts its SSID
- D. Transmit a connect message on the station's published frequency

To make contact, one sends a connection message on the published frequency, as in **Answer D**. The other choices will not work.

G2E11 Which of the following is characteristic of the FT8 mode of the WSJT-X family?
- A. It is a keyboard-to-keyboard chat mode
- B. Each transmission takes exactly 60 seconds
- C. It is limited to use on VHF
- D. Typical exchanges are limited to call signs, grid locators, and signal reports

Not all digital modes support long QSOs. In particular, FT8 sends the minimum information: call signs, grid locations, and signal reports as Figure 2.4 shows. This example makes **Answer D** the correct choice. Each of the other answers is an incorrect statement about FT8. See https://physics.princeton.edu//pulsar/K1JT/wsjtx.html for more information about the WSJT-X modes.

G2E12 Which of the following connectors would be a good choice for a serial data port?
- A. PL-259
- B. Type N
- C. Type SMA
- D. DE-9

The connectors listed in Answers A, B, and C are all types of Radio Frequency (RF) connectors, so they do not make for good data connectors. The DE-9 connector of **Answer D** is a data connector, so it is the right choice. Chapter 6 has more about typical connectors.

G2E13 Which communication system sometimes uses the internet to transfer messages?
  A. Winlink
  B. RTTY
  C. ARES
  D. SKYWARN

The Winlink protocol can use the Internet to transfer messages, so **Answer A** is the right choice. The other protocols use only RF communications, so they are incorrect.

G2E14 What could be wrong if you cannot decode an RTTY or other FSK signal even though it is apparently tuned in properly?
  A. The mark and space frequencies may be reversed
  B. You may have selected the wrong baud rate
  C. You may be listening on the wrong sideband
  D. All of these choices are correct

Each of the fault modes in Answers A, B, and C can cause this condition. These conditions make **Answer D** the best choice for this question.

G2E15 Which of the following is a requirement when using the FT8 digital mode?
  A. A special hardware modem
  B. Computer time accurate within approximately 1 second
  C. Receiver attenuator set to -12 dB
  D. A vertically polarized antenna

In Figure 2.3, we can see that the software divides the waterfall display into time segments and that the transmissions align with the start and stop of these segments. In FT8, the operators must keep their transmissions within these timing windows by having accurate clocks synchronized with a timing protocol. This requirement makes **Answer B** the correct choice. The other options are not requirements for FT8 operation.

Chapter 3

G3 — RADIO WAVE PROPAGATION

3.1 Introduction

In the Technician Class study guide, we began looking into how radio waves move through the Earth's atmosphere. In addition to the layers of the Earth's atmosphere like the troposphere, where we live and our weather occurs, the ionosphere is important in reflecting radio waves and supporting long-distance communications on the High Frequency (HF) bands. The General Class license examination has always had a strong emphasis on the HF radio propagation effects that an operator having multi-mode, worldwide access needs to understand. For the General Class examination, we need to know much more about the Sun and its effects on the Earth's atmosphere, especially the Earth's ionosphere.

The *Radio Wave Propagation* subelement has the following question groups:

A. Solar Activity
B. HF Propagation
C. Ionospheric Characteristics

Subelement 3 will generate three questions on the General Class examination.

3.2 Radio Engineering Concepts

3.2.1 Solar Activity

Sunspot Cycles The first solar activity we need to understand is the sunspot cycle. On the Sun's surface, regional magnetic storms called *sunspots* appear and disappear just as weather patterns do on the Earth. An individual spot, like the large ones in Figure 3.1, can last from a few days to a few months, and they correlate with the amount of radiation hitting the Earth's ionosphere. Scientists

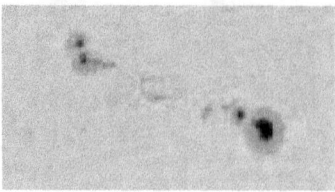

Figure 3.1: A sunspot group. Image courtesy NOAA.

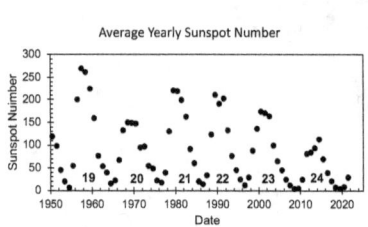

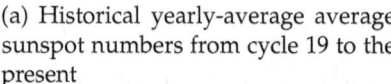

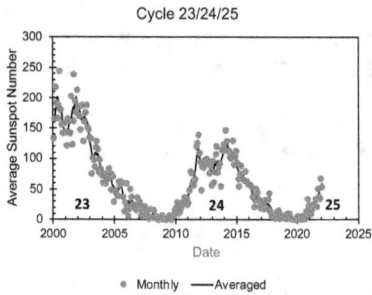

(a) Historical yearly-average average sunspot numbers from cycle 19 to the present

(b) Sunspot monthly activity for cycles 23 and 24, and the start of cycle 25

Figure 3.2: Solar activity as measured by sunspot numbers. Data courtesy WDC-SILSO, Royal Observatory of Belgium, Brussels

measure these sunspot patterns by the sunspot number, which has an 11-year cycle, as Figure 3.2 illustrates. Part (a) of the figure shows the historical sunspot number data for the past few cycles, where you can see the 11-year period. Part (b) of the figure shows we are currently in the early stages of sunspot cycle 25. The cycle 24 sunspot maximum was in 2014, and its sunspot minimum was in the spring of 2020. Notice that the cycle 24 maximum was a double maximum, with one peak in 2011/2012 and the second in 2014. Also, notice that the last maximum was much lower than the recent historical data for maximums. While cycle 24 officially ended in December 2019, it was still producing sunspots through the spring of 2020. Cycle 25 began in December 2019, and scientists expect the sunspot maximum to occur around 2025. What are the implications of this pattern? Generally, the HF bands will have better propagation conditions near the maximum because they will be interacting with more charged particles from the Sun when there are more sunspots. This condition does not mean that HF propagation does not happen during a solar minimum. Instead, the propagation quality is lower, and long-distance propagation does not occur as frequently.

While the sunspot number correlates with the amount of radiation hitting the ionosphere, there are more direct measures that scientists use to keep track of the radiation. One measure is the *solar flux index* which is a measure of solar noise at 10.7 cm (2800 MHz).

Solar Disturbances Disturbances from the Sun correlate with sunspot activity. This correlation does not mean that a solar disturbance cannot occur at a sunspot minimum. It is just less likely then. During a solar disturbance, the Sun emits electromagnetic radiation and particles that can strike the Earth. The electromagnetic radiation arrives within 8 minutes because it travels at the speed of light. Particles arrive many hours later, frequently 1 to 2 days later, because they are not moving at the speed of light. Figure 3.3 shows solar activity affecting radio propagation that includes

Solar Flares — an eruption on the surface of the sun, usually linked with a sunspot, sending large numbers of protons and electromagnetic radiation outward; flares arise in a few minutes and may last for several hours

Coronal Mass Ejection — a large bubble of material ejected from the solar corona (solar atmosphere) over several hours and follows the solar magnetic field lines; it can be associated with a flare but can occur without a flare as well

Coronal Holes — appear as large, dark openings in the corona and produce a high-speed solar wind

The rapid bombardment of the Earth's ionospheric layer by the emissions from a flare can cause a Sudden Ionospheric Disturbance (SID) in the ionosphere.

Geomagnetic Activity The charged particles emitted by the Sun will interact with the Earth's magnetic field and cause a *geomagnetic storm*. Scientists measure these interactions to indicate the strength of the *space weather* caused by the Sun. While the geomagnetic indices do not measure the ionosphere, the results correlate with the presence of ionospheric activity, and they can indicate when radio propagation may be adversely affected. The measures we need to know are

K Index — a magnetometer measurement of the Earth's magnetic field relative to the "normal" value, averaged every 3 hours made at several sites around the Earth, with the result reported on a logarithmic scale from 0 to 9

A Index — a linear-scale measurement derived from the K Index made at sites around the Earth and averaged over the day with values ranging from 0 to 400

The Space Weather Radio Communications Dashboard Web site `https://www.swpc.noaa.gov/communities/radio-communications` posts current measures of important propagation-related data from National Oceanic and Atmospheric Administration (NOAA).

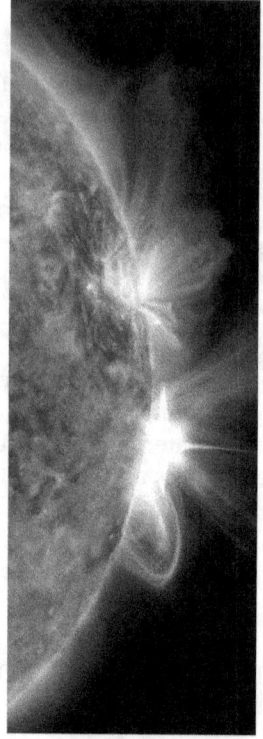

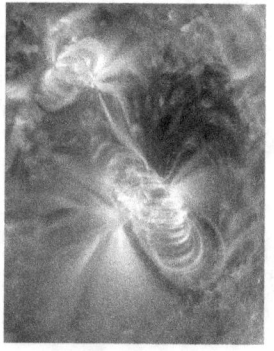

(a) Image of a solar flare.

(b) Image of a coronal mass ejection.

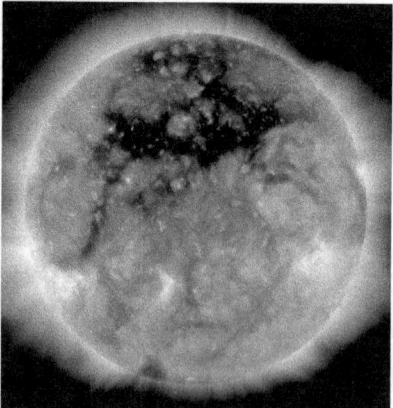

(c) Image of a coronal hole.

Figure 3.3: Solar events that affect the ionosphere and radio propagation. All images courtesy of NASA

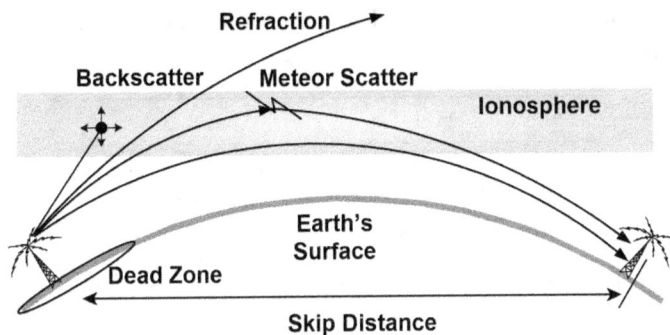

Figure 3.4: Various forms of skywave propagation between stations.

3.2.2 Radio Propagation

During our Technician Class examination studies, we referred to Figure 3.4 for the multiple types of skywave propagation. For the General Class examination, we need to understand more of the details of this diagram. The first concept is the *short path* versus the *long path* propagation that we saw in Figure 2.5. With a short path, the signal bounces off the ionosphere at least once while connecting the transmitter and the receiver. However, the signal may also make multiple bounces off the ionosphere and the Earth's surface, thereby taking the long path around the Earth to arrive at the receiver. Interestingly, the receiver may hear both signals, but with a slight delay between them.

The second concept comes from examining the two rays where one is bent and goes outside the ionosphere, while the second is bent and comes back to Earth. All other variables being equal, we can see this as a demonstration of the frequency-dependent characteristics of the ionosphere. The Maximum Usable Frequency (MUF) indicates the maximum frequency that can sustain a communications link between two users. Above that frequency, the signal will not return to the Earth but go out through the ionosphere. Similarly, the Lowest Usable Frequency (LUF) is the lowest frequency that will sustain a communications link between two users. Below the LUF, the ionosphere generally absorbs the signal keeping the link from operating.

As Figure 3.4 shows, the ionosphere acts as a mirror reflecting the signals between the LUF and the MUF back to the Earth. Because the ionosphere has different layers at different heights, the reflection point will also be at different altitudes in the atmosphere.

You can subscribe to periodic radio propagation bulletins, such as the ones available from ARRL, at http://www.arrl.org/bulletins. This bulletin covers sunspot numbers, geomagnetic values, and other useful information in a form useful for amateur radio operators.

Table 3.1: Approximate One-Hop Propagation Distances for Different Ionospheric Layers.

Layer	Height	Distance
D	60 – 90 km	absorbs HF signals
E	100 – 125 km	1250 miles
F1	300 km	2000 miles
F2	400 km	2500 miles

3.2.3   The Ionosphere

Structure   As we see in Table 3.1, the ionosphere has several main layers. These layers result from the Sun's emitting particles and radiation that hits the Earth. The F-layer has two sub-layers: the F1 and F2 layers. The F layers show separation into different concentrations during daylight but collapse back into a single layer at night. All the layers are stronger during the day than at night.

Critical Frequency   One technique scientists use to measure the state of the ionosphere is to direct a Radio Frequency (RF) wave vertically. They vary the frequency until a given ionospheric layer stops returning the wave and, instead, the wave passes to the next higher layer or out of the ionosphere. They build up a profile of the *critical frequency* of where this reflection stops for each layer to form an *ionogram* for the atmosphere.

Critical Angle   The absorption for each layer of the ionosphere depends on the angle the radio signal makes with the layer as it interacts with it. The *critical angle* is the highest takeoff angle (measured up from the horizontal) that will return a signal to the Earth.

Skip Zone and Skip Distance   In Figure 3.4 we saw the *skip distance* illustrated as the distance between the transmitter and the receiver along the Earth's surface when the ionosphere reflects the signal. As the distance gets smaller, there becomes a spacing where the ionosphere does not reflect the RF signal into the region, and communication does not occur. This spacing forms the *skip zone* or the *dead zone* in the figure because communications cannot reach that shorter distance than the skip distance.

Scatter   Figure 3.4 also shows scatter propagation, which is mostly an E-layer phenomenon. With scatter, small charged regions form and act like a mosaic mirror that reflects the signals randomly into multiple directions. Sometimes, the scatter can reflect signals into the skip zone. The multiple-scatter nature of the region often distorts the scatter signals.

3.3 G3A - Solar Activity

3.3.1 Overview

The *Solar Activity* question group in Subelement G3 tests your understanding of solar activities that have an important influence on radio wave propagation on Earth. The *Solar Activity* group covers topics such as
- Sunspots and solar radiation
- Ionospheric disturbances
- Propagation forecasting and indices

The test producer will select one of the 14 questions in this group for your exam.

3.3.2 Questions

G3A01 What is the significance of the sunspot number with regard to HF propagation?
  A. Higher sunspot numbers generally indicate a greater probability of good propagation at higher frequencies
  B. Lower sunspot numbers generally indicate greater probability of sporadic E propagation
  C. A zero sunspot number indicate radio propagation is not possible on any band
  D. A zero sunspot number indicates undisturbed conditions

Sunspots, like those in Figure 3.1, affect the Earth's ionosphere, so they also affect HF propagation. The general rule of thumb is that the higher the number of sunspots, the better the HF propagation conditions are on average, as in **Answer A**. With low sunspot numbers, sporadic-E propagation is less likely. Answer C is not true; for example, the 20-m and lower bands propagate RF even at low sunspot numbers. Answer D is a distraction because the Sun still produces some radiation that affects propagation when there are no sunspots.

G3A02 What effect does a Sudden Ionospheric Disturbance have on the day-time ionospheric propagation of HF radio waves?
  A. It enhances propagation on all HF frequencies
  B. It disrupts signals on lower frequencies more than those on higher frequencies
  C. It disrupts communications via satellite more than direct communications
  D. None, because only areas on the night side of the Earth are affected

During a SID, lower frequencies are affected more than higher frequencies by ionospheric interference, as in **Answer B**. Because of Answer B, all frequencies are not affected equally. The SID affects any communications interacting with the ionosphere, so if the contacts are on the ground or from space, it does not matter. Answer D is silly since the night side of the Earth faces away from the direction from which the charged particles from the Sun are coming.

G3A03 Approximately how long does it take the increased ultraviolet and X-ray radiation from solar flares to affect radio propagation on the Earth?
  A. 28 days
  B. 1 to 2 hours
  C. 8 minutes
  D. 20 to 40 hours

Solar flares, like the one in Figure 3.3a, eject energetic particles and photons from the sun. This question deals with different types of photon radiation that travel at the speed of light. The light travel time from the Sun to the Earth is 8 minutes, so the correct choice is **Answer C**. Answer A is the Sun's rotation period. Answer D may sound familiar because it is the time for charged particles emitted by the Sun to reach the Earth, but not electromagnetic radiation. Answer B is a distraction.

G3A04 Which of the following are least reliable for long distance communications during periods of low solar activity?
  A. 80 meters and 160 meters
  B. 60 meters and 40 meters
  C. 30 meters and 20 meters
  D. 15 meters, 12 meters and 10 meters

This question will make better sense to you once you have operating experience on the HF bands. Generally, low solar activity periods regularly have impaired conditions on the bands above 20 m. Another way of saying that is the bands are less reliable above 14 MHz. Since all the choices except Answer D are below the 20-m band, and the question is looking for the least reliable communications, the correct choice is the frequency list in **Answer D**. The other frequencies will be able to sustain some level of communications even with low solar activity.

G3A05 What is the solar flux index?
  A. A measure of the highest frequency that is useful for ionospheric propagation between two points on the Earth
  B. A count of sunspots which is adjusted for solar emissions
  C. Another name for the American sunspot number
  D. A measure of solar radiation at 10.7 centimeters wavelength

The solar flux index is a measurement of the solar radiation at 10.7 cm, so the correct choice is **Answer D**. Answer A is the Maximum Usable Frequency (MUF), while Answers C is another name for the sunspot number produced by NOAA. Answer B is technobabble.

G3A06 What is a geomagnetic storm?
  A. A sudden drop in the solar flux index
  B. A thunderstorm that affects radio propagation
  C. Ripples in the ionosphere
  D. A temporary disturbance in the Earth's magnetosphere

The key here is the word "geomagnetic" in the question statement. That means the Earth's magnetic field is involved. **Answer D** is the only one that deals with the Earth's magnetic field, so it is the right choice. Answer A would be good to know for propagation prediction, but it is a distraction here. Thunderstorm radio propagation effects are primarily for frequencies above 10 GHz, where the rain attenuates the signal. However, a thunderstorm is still not a geomagnetic storm. Answer C is to distract you as well.

G3A07 At what point in the solar cycle does the 20-meter band usually support worldwide propagation during daylight hours?
  A. At the summer solstice
  B. Only at the maximum point of the solar cycle
  C. Only at the minimum point of the solar cycle
  D. At any point in the solar cycle

Again, this is an operating experience question. The 20-m band is generally "open" during daylight hours at any point in the solar cycle. Therefore, the correct choice is **Answer D**. The other options are to distract you.

G3A08 Which of the following effects can a geomagnetic storm have on radio propagation?
  A. Improved high-latitude HF propagation
  B. Degraded high-latitude HF propagation
  C. Improved ground wave propagation
  D. Degraded ground wave propagation

So far, we have seen that geomagnetic storms generally make radio propagation worse. Therefore, Answers A and C are wrong since they imply improved communications in some manner. The correct choice is **Answer B** since it is the only one indicating degraded communications on the HF bands. Answer D is incorrect because it deals with ground wave propagation, not atmospheric propagation, where the geomagnetic storm causes interactions.

G3A09 What benefit can high geomagnetic activity have on radio communications?
  A. Auroras that can reflect VHF signals
  B. Higher signal strength for HF signals passing through the polar regions
  C. Improved HF long path propagation
  D. Reduced long delayed echoes

Geomagnetic storms can increase auroral activity, and these auroras can reflect Very High Frequency (VHF) transmissions, so **Answer A** is the right choice. Answers B, C, and D do not match the way propagation works under these conditions.

G3A10 What causes HF propagation conditions to vary periodically in a 28-day cycle?
  A. Long term oscillations in the upper atmosphere
  B. Cyclic variation in the Earth's radiation belts
  C. The sun's rotation on its axis
  D. The position of the moon in its orbit

The Sun rotates once every 28 days, so the solar activity has a 28-day repetition component. This rotation makes **Answer C** the correct choice for answering this question. The other options are incorrect statements to distract you.

G3A11 How long does it take charged particles from coronal mass ejections to affect radio propagation on Earth?
  A. 28 days
  B. 14 days
  C. 4 to 8 minutes
  D. 20 to 40 hours

Charged particles move quickly, but not at the speed of light like photons do. They take 20 to 40 hours to reach the Earth, making **Answer D** the correct choice. The solar rotation rate is 28 days making Answer A incorrect. The other options are to distract you.

G3A12 What does the K-index indicate?
  A. The relative position of sunspots on the surface of the sun
  B. The short-term stability of Earth's magnetic field
  C. The stability of the sun's magnetic field
  D. The solar radio flux at Boulder, Colorado

There are two indices of geomagnetic activity, so we have two questions about them. The first is the K-index, which measures the geomagnetic activity over three hours. This measurement period makes **Answer B** the correct choice for this question. Since this is a geomagnetic index, the K-index does not measure solar parameters as in the other choices.

G3A13 What does the A-index indicate?
  A. The relative position of sunspots on the surface of the sun
  B. The amount of polarization of the sun's electric field
  C. The long-term stability of Earth's geomagnetic field
  D. The solar radio flux at Boulder, Colorado

This question is about the second geomagnetic index, the A-index. The A-index is the daily average based on the K-index and ranges from 0 through 400. **Answer C** is the correct choice among those given. Since this is a geomagnetic index, the A-index does not measure solar parameters as in the other choices.

G3A14 How are radio communications usually affected by the charged particles that reach Earth from solar coronal holes?
   A. HF communications are improved
   B. HF communications are disturbed
   C. VHF/UHF ducting is improved
   D. VHF/UHF ducting is disturbed

A coronal hole is associated with the Sun emitting a burst of charged particles. The charged particles interact with the Earth's magnetic field, which causes disruptions to HF communications. This interaction makes **Answer B** the right choice for this question. Answer A is incorrect because the HF is not improved. VHF or Ultra High Frequency (UHF) ducting is a weather effect in the troposphere, so Answers C and D are incorrect.

3.4   G3B - HF Propagation

3.4.1   Overview

The *HF Propagation* question group in Subelement G3 quizzes you on concepts dealing with radio wave propagation through the Earth's atmosphere. The *HF Propagation* group covers topics such as
   • Maximum Usable Frequency
   • Lowest Usable Frequency
   • Propagation
The test producer will select one of the 11 questions in this group for your exam.

3.4.2   Questions

G3B01 What is a characteristic of skywave signals arriving at your location by both short-path and long-path propagation?
   A. Periodic fading approximately every 10 seconds
   B. Signal strength increased by 3 dB
   C. The signal might be cancelled causing severe attenuation
   D. A slightly delayed echo might be heard

In this case, you will hear the long-path signal as a time-delayed version of the short-path signal, which makes **Answer D** the right choice for the question. The distraction choices in Answers A, B, and C are to see if you can be confused by technical-sounding answers.

G3B02 What factors affect the MUF?
A. Path distance and location
B. Time of day and season
C. Solar radiation and ionospheric disturbances
D. All these choices are correct

Each of the factors listed in Answers A, B, and C affects the ionosphere's MUF, so **Answer D** is the best choice.

G3B03 Which of the following applies when selecting a frequency for lowest attenuation when transmitting on HF?
A. Select a frequency just below the MUF
B. Select a frequency just above the LUF
C. Select a frequency just below the critical frequency
D. Select a frequency just above the critical frequency

Here we need to remember that the ionosphere will not permit propagation at all frequencies under all conditions. When establishing a link between two stations using a reflection off the ionosphere, there will be a maximum frequency that can support this link. Transmitting a radio signal just below the MUF frequency is the best option among those given here, so **Answer A** is the best choice. When you use a frequency just above the LUF, you may still encounter some absorption, so this option is not as good as Answer A. "Critical frequencies" are slightly different in propagation. They are the frequencies above which the radio signal does not reflect but passes through the ionosphere. These frequencies are different than what the question is asking about.

G3B04 What is a reliable way to determine if the MUF is high enough to support skip propagation between your station and a distant location on frequencies between 14 and 30 MHz?
A. Listen for signals from an international beacon in the frequency range you plan to use
B. Send a series of dots on the band and listen for echoes from your signal
C. Check the strength of TV signals from western Europe
D. Check the strength of signals in the MF AM broadcast band

This question is asking about HF propagation conditions, so we can eliminate Answers C and D because they involve Medium Frequency (MF) and VHF signals. Answer B is a silly distractor. Listening for beacon stations, like the ones mentioned in Chapter 1, in your band of interest can help you determine if the band is open, so **Answer A** is the best choice among those given.

G3B05 What usually happens to radio waves with frequencies below the MUF and above the LUF when they are sent into the ionosphere?
  A. They are bent back to the Earth
  B. They pass through the ionosphere
  C. They are amplified by interaction with the ionosphere
  D. They are bent and trapped in the ionosphere to circle the Earth

Operating above the LUF and below the MUF is a desirable condition. In that case, the radio signal will be sent back to the Earth by the ionosphere, which enables long-distance communications. **Answer A** is the correct choice for this question. Answers B and C are technically incorrect. Answer D is a silly distraction.

G3B06 What usually happens to radio waves with frequencies below the LUF?
  A. They are bent back to the Earth
  B. They pass through the ionosphere
  C. They are completely absorbed by the ionosphere
  D. They are bent and trapped in the ionosphere to circle the Earth

Frequencies below the LUF are absorbed, as in **Answer C**. Answers A and B deal with frequencies on either side of the MUF. Answer D is still humorous.

G3B07 What does LUF stand for?
  A. The Lowest Usable Frequency for communications between two points
  B. The Longest Universal Function for communications between two points
  C. The Lowest Usable Frequency during a 24-hour period
  D. The Longest Universal Function during a 24-hour period

As we saw in the previous questions, LUF is the Lowest Usable Frequency between points, so **Answer A** is the correct choice. Be careful when taking the test because the other options look similar. The LUF is an instantaneous measurement and not averaged over 24 hours.

G3B08 What does MUF stand for?
  A. The Minimum Usable Frequency for communications between two points
  B. The Maximum Usable Frequency for communications between two points
  C. The Minimum Usable Frequency during a 24-hour period
  D. The Maximum Usable Frequency during a 24-hour period

This question is asking about a term we saw earlier. The MUF is the Maximum Usable Frequency between points as in **Answer B**. As with the LUF it is an instantaneous measurement and not a 24-hour average. Be careful with the distraction answers.

G3B09 What is the approximate maximum distance along the Earth's surface that is normally covered in one hop using the F2 region?
- A. 180 miles
- B. 1,200 miles
- C. 2,500 miles
- D. 12,000 miles

The F2 region is the highest region of concern in the ionosphere, so we use it to return the signals to the Earth. As Table 3.1 shows, the correct choice is **Answer C**. Answers A is for tropospheric propagation. The 12,000 miles in Answer D is for moon bounce, while 1,200 miles is for the E region.

G3B10 What is the approximate maximum distance along the Earth's surface that is normally covered in one hop using the E region?
- A. 180 miles
- B. 1,200 miles
- C. 2,500 miles
- D. 12,000 miles

The E region is lower in the ionosphere than the F region but above the D region. Therefore, Answer C is out since that answer goes with the F region. As Table 3.1 shows, the 1,200 miles of **Answer B** is the correct choice. Answer A is too short for the path's geometry. Answer D is to distract you.

G3B11 What happens to HF propagation when the LUF exceeds the MUF?
- A. No HF radio frequency will support ordinary skywave communications over the path
- B. HF communications over the path are enhanced
- C. Double hop propagation along the path is more common
- D. Propagation over the path on all HF frequencies is enhanced

By reasoning through the question statement, you should be able to see that the HF bands will be dead, as far as the operator is concerned, since no HF skywave propagation will occur. This condition makes **Answer A** the correct choice for the question. Answer B is the opposite, so it is not a good choice. Answers C and D sound good, but they are just technobabble.

3.5   G3C - Ionospheric Characteristics

3.5.1   Overview

The *Ionospheric Characteristics* question group in Subelement G3 covers ionospheric effects encountered in radio wave propagation on the amateur bands. The *Ionospheric Characteristics* group covers topics such as

- Ionospheric layers
- Critical angle and frequency
- HF scatter
- Near Vertical Incidence Skywave

The test producer will select one of the 11 questions in this group for your exam.

3.5.2 Questions

G3C01 Which ionospheric layer is closest to the surface of the Earth?
A. The D layer
B. The E layer
C. The F1 layer
D. The F2 layer

If you go back to Table 3.1 on atmospheric layers, you will see that the D layer is closest to the Earth, then the E layer, and then the two F layers. From this, **Answer A** is the correct response.

G3C02 Where on the Earth do ionospheric layers reach their maximum height?
A. Where the Sun is overhead
B. Where the Sun is on the opposite side of the Earth
C. Where the Sun is rising
D. Where the Sun has just set

The F region reacts to ultraviolet radiation from the Sun. This radiation's maximum intensity occurs at noon during the summer when the Sun is highest in the sky at your location. Therefore, the best choice among those given here is when the Sun is overhead, as in **Answer A**. The other options do not correspond to local noon, so they are incorrect choices.

G3C03 Why is the F2 region mainly responsible for the longest distance radio wave propagation?
A. Because it is the densest ionospheric layer
B. Because of the Doppler effect
C. Because it is the highest ionospheric region
D. Because of meteor trails at that level

As we saw earlier, the F region is highest in the atmosphere, so the correct response is **Answer C**. Answer A is technically incorrect, while the Doppler shift in Answer B comes from transmitter or receiver motions. The meteor trails in Answer D tend to be in the E layer, which is below the F layer.

G3C04 What does the term "critical angle" mean, as used in radio wave propagation?
- A. The long path azimuth of a distant station
- B. The short path azimuth of a distant station
- C. The lowest takeoff angle that will return a radio wave to the Earth under specific ionospheric conditions
- D. The highest takeoff angle that will return a radio wave to the Earth under specific ionospheric conditions

The "critical angle" definition is given in **Answer D**, so that is the right choice. Answer C is the opposite angle, so it is incorrect. Answers A and B are to distract you.

G3C05 Why is long distance communication on the 40-meter, 60-meter, 80-meter and 160-meter bands more difficult during the day?
- A. The F layer absorbs signals at these frequencies during daylight hours
- B. The F layer is unstable during daylight hours
- C. The D layer absorbs signals at these frequencies during daylight hours
- D. The E layer is unstable during daylight hours

The D region is lowest in the atmosphere, and it tends to absorb long-wavelength radiation. Therefore, **Answer C** is the correct choice. Answers A, B, and D are technically incorrect, so they are just there to distract you.

G3C06 What is a characteristic of HF scatter signals?
- A. Phone signals have high intelligibility
- B. Signals have a fluttering sound
- C. There are very large, sudden swings in signal strength
- D. Scatter propagation occurs only at night

HF scatter is caused by radio waves being bounced back to the Earth by small charged regions in the ionosphere. In this case, the ionosphere is acting like a bumpy mirror. Therefore, a distorted signal is received. Among the choices given to describe this effect, **Answer B** is the best one. Answer A is just the opposite and wrong. Answer C is incorrect because the signal strength may not have a large signal swing. Scatter propagation tends to be more common during the day than at night.

G3C07 What makes HF scatter signals often sound distorted?
- A. The ionospheric layer involved is unstable
- B. Ground waves are absorbing much of the signal
- C. The E-region is not present
- D. Energy is scattered into the skip zone through several different radio wave paths

Answers B and C are technically incorrect, so they are eliminated. Energy scattering, as in **Answer D**, is the correct choice to answer the question. Answer A may sound plausible, but it is not as good a description as Answer D.

G3C08 Why are HF scatter signals in the skip zone usually weak?
A. Only a small part of the signal energy is scattered into the skip zone
B. Signals are scattered from the magnetosphere, which is not a good reflector
C. Propagation is through ground waves, which absorb most of the signal energy
D. Propagation is through ducts in the F region, which absorb most of the energy

HF scatter is an E-region process in the ionosphere, so Answer D is wrong because it deals with the F region. Since it is an ionospheric process, there are no ground waves involved, and we can eliminate Answer C. Answer B sounds technical, but it is not part of radio propagation, so this is not a good choice. The scattering process only involves a small amount of the radio energy entering the skip zone, so the correct choice to answer this question is **Answer A**.

G3C09 What type of propagation allows signals to be heard in the transmitting station's skip zone?
A. Faraday rotation
B. Scatter
C. Chordal hop
D. Short-path

This question is another way of asking about the radio scatter process, so the correct choice is **Answer B**. Answer A is incorrect because it deals with radio waves coming in from satellites. Answers C and D are propagation modes, but not the kind here, so they are wrong.

G3C10 What is Near Vertical Incidence Skywave (NVIS) propagation?
A. Propagation near the MUF
B. Short distance MF or HF propagation using high elevation angles
C. Long path HF propagation at sunrise and sunset
D. Double hop propagation near the LUF

Near Vertical Incidence Skywave (NVIS) is a technique where the operator transmits a radio signal almost straight up (high elevation angle) rather than the typical HF approach of sending the radio signals close to the horizon. The radio signals interact with the ionosphere and return to Earth at a relatively short distance from the transmitter, approximately 200 miles. This procedure makes **Answer B** the correct choice. The other options are technobabble distractions to see if you know what NVIS is.

G3C11 Which ionospheric layer is the most absorbent of long skip signals during daylight hours on frequencies below 10 MHz?
- A. The F2 layer
- B. The F1 layer
- C. The E layer
- D. The D layer

Here you need to remember that the D layer is the most absorbent during daylight hours, so the correct choice is **Answer D**. The others are above the D layer in altitude, so their effects are not as pronounced.

Chapter 4

G4 — AMATEUR RADIO PRACTICES

4.1 Introduction

In our Technician Class license studies, we saw the basic operations of radios in a typical amateur's "radio shack." The General Class license exam will involve you more with the details of operating the equipment, especially for the High Frequency (HF) bands. The *Amateur Radio Practices* subelement has the following question groups:

A. Station Operation
B. Testing
C. Interference
D. Transceiver Operations
E. Remotely Operating

Subelement 4 will generate five questions on the General Class examination.

4.2 Radio Engineering Concepts

Shack Configuration   In our Technician studies, we saw Figure 4.1 to illustrate an overview of a typical amateur's radio shack. Figure 4.1 illustrates the transceiver and some of the typical auxiliary equipment to support the operations. This can include a computer, one or more antennas, power supply, modem, and other devices.

Rig Controls   The General Class examination will ask about many control functions for a typical HF Transceiver (XCVR). To help with this, Figure 4.2 illustrates a commercial HF rig of the type found in many amateur shacks. Important controls for the General Class examination include

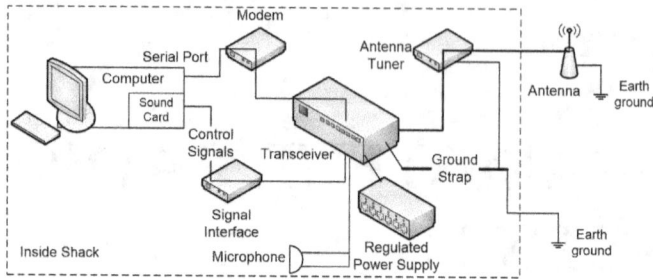

Figure 4.1: General amateur radio shack configuration for HF operations.

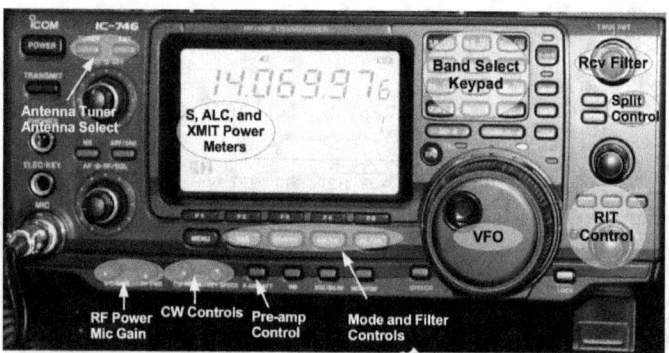

Figure 4.2: Typical controls for a commercial HF amateur transceiver.

Antenna Tuner — engage and tune the internal antenna tuner (many XCVRs have multiple antenna ports)

Antenna Select — choose between available antennas

ALC Meter — show the current Automatic Level Control (ALC) level

Power Meter — show the current output power level

RF Power Set — RF output power level adjustment knob

Mic Gain — audio input level set knob

CW Controls — knobs to set the internal Continuous Wave (CW) tone frequency and keying speed

Pre-amp Control — select the level of signal pre-amplification in the receiver

Mode and Filtering Control — set the transmission mode (sideband, CW/RTTY, and AM/FM) and filtering characteristics

Band Select Keypad — set the desired operating band (each pad key has multiple settings for frequency and mode)

Rcv Filter — set the internal receiver filter to better isolate the desired transmission on the band

Split Control — control the ability to transmit and receive on different frequencies

RIT Control — control the Receiver Incremental Tuning (RIT) operating characteristics

VFO — Variable Frequency Oscillator (VFO) tuning control knob

Other rigs, especially the newer models will have additional controls.

Filters   Filters are electronic circuits that operate in the frequency domain, as in Figure 4.3. The filter's design removes signal components in a specific frequency region. The filter does this by attenuating the desired signal components and passing the other signal components with little or no attenuation outside the specified frequency region Filters are named based on their passband, which has unattenuated signal components. Figure 4.3 shows these characteristics. Common filters are

High Pass Filter (HPF) — passes all frequencies above a specified cutoff frequency

Low Pass Filter (LPF) — passes all frequencies below a specified cutoff frequency

Band Pass Filter (BPF) — passes all frequencies between a specified high and low cutoff frequency

Band Stop — passes all frequencies except those between a specified high and low cutoff frequency

We will see more about filters in Chapter 7.

RF Amplifiers   Figure 4.4 shows that the XCVR has two radio components that share an antenna: a transmitter and a receiver. Typically, an internal switch alternates the antenna connection between the two modules. Figure 4.4 shows that the XCVR has two amplifiers inside the rig. The Radio Frequency (RF)

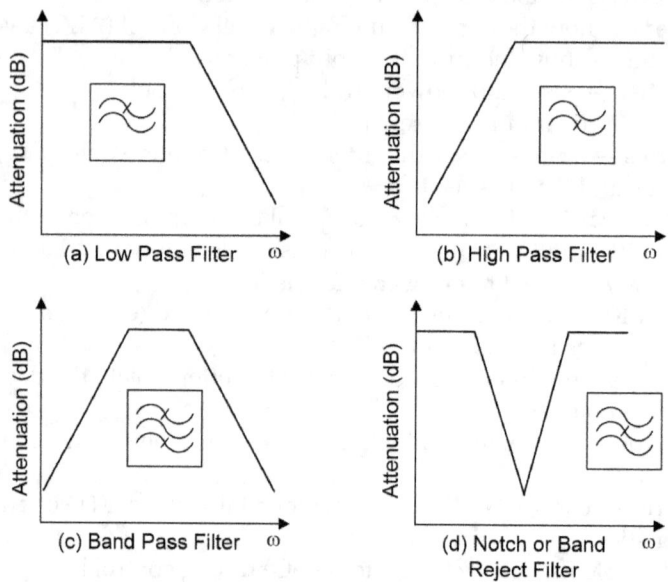

Figure 4.3: Filter frequency domain operations and their associated circuit symbols.

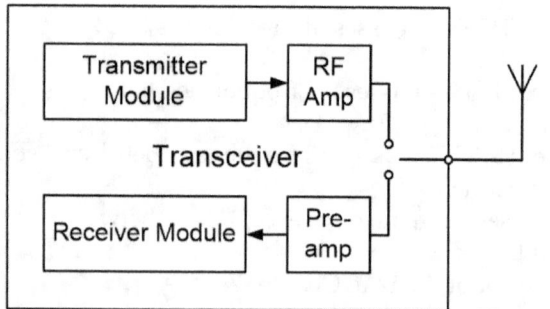

Figure 4.4: General block diagram for a HF transceiver.

amplifier boosts the signal going to the antenna. Typical commercial XCVRs can produce up to 100 W output power "out of the box." The pre-amp amplifies the received signal before sending it to the demodulator. Both amplifiers are under user control in the XCVR configuration settings.

Many XCVRs also have external RF amplifiers to make high-powered signals beyond the rig's internal settings. There are two general classes of external amplifiers: *tube amplifiers* and *solid-state amplifiers*. The tube amplifiers operate by changing the currents on the tube's electrical plates and generating higher output power levels than the solid-state amplifiers. Solid-state amplifiers use RF power transistors to amplify the signal.

4.3 G4A - Station Operation

4.3.1 Overview

The *Station Operation* question group in Subelement G4 tests you on operational characteristics of HF rigs and amplifiers. The test producer will select one of the 17 questions in this group for your exam.

4.3.2 Questions

G4A01 What is the purpose of the "notch filter" found on many HF transceivers?
   A. To restrict the transmitter voice bandwidth
   B. To reduce interference from carriers in the receiver passband
   C. To eliminate receiver interference from impulse noise sources
   D. To enhance the reception of a specific frequency on a crowded band

A notch filter, also known as a band stop filter, passes all frequencies except for those in a very narrow region (the "notch"), as Figure 4.3 shows. Figure 6.1 includes the circuit symbol for a band stop filter. With proper notch placement, the radio can minimize the effects of nearby carriers in the receiver when trying to capture a signal of interest. This property makes **Answer B** the right choice. The notch eliminates a specific frequency, not enhances it, so Answer D is incorrect. Impulse noise is wideband, so a notch filter will not remove all the noise. This question is not about a transmitter filter, so Answer A is incorrect.

G4A02 What is one advantage of selecting the opposite, or "reverse," sideband when receiving CW signals on a typical HF transceiver?
   A. Interference from impulse noise will be eliminated
   B. More stations can be accommodated within a given signal passband
   C. It may be possible to reduce or eliminate interference from other signals
   D. Accidental out of band operation can be prevented

Impulse noise is wideband, and it will interfere on both sidebands. The bandwidth is the same on both sidebands, and more stations cannot use the band in either mode. Unless you are using bad amateur practice and operating on precisely the band's edge, staying with either sideband will keep you in the band and not prevent accidentally going outside of the band. The correct answer is that you can sometimes reduce or eliminate interference during CW QSOs by using this operating trick, so **Answer C** is the right choice.

G4A03 What is normally meant by operating a transceiver in "split" mode?
   A. The radio is operating at half power
   B. The transceiver is operating from an external power source
   C. The transceiver is set to different transmit and receive frequencies
   D. The transmitter is emitting an SSB signal, as opposed to DSB operation

Split mode means that you are operating at different transmit and receive frequencies to assist in working with other stations. This configuration makes **Answer C** the right choice. Once you have experience operating on the HF bands, you will see that the other choices are silly distractions.

G4A04 What reading on the plate current meter of a vacuum tube RF power amplifier indicates correct adjustment of the plate tuning control?
   A. A pronounced peak
   B. A pronounced dip
   C. No change will be observed
   D. A slow, rhythmic oscillation

This question is asking about how a "dip meter" works, so the correct choice is **Answer B**. Answer A is the opposite effect, although it may sound like the right choice. Answers C and D are to distract you.

G4A05 What is a reason to use Automatic Level Control (ALC) with an RF power amplifier?
   A. To balance the transmitter audio frequency response
   B. To reduce harmonic radiation
   C. To reduce distortion due to excessive drive
   D. To increase overall efficiency

Proper use of an ALC circuit will help you maintain a constant output level, thereby reducing distortion from an excessive amplifier input level, so **Answer C** is the right choice. Answer D may look correct, but it does not explain things fully. Answers A and B would be valuable functions, but they are not what an ALC circuit does.

G4A06 What type of device is often used to match transmitter output impedance to an impedance not equal to 50 ohms?
A. Balanced modulator
B. SWR Bridge
C. Antenna coupler or antenna tuner
D. Q Multiplier

As we saw in the Technician Class study guide, impedance matching is essential. The antenna coupler or an antenna tuner are devices to make that happen, so **Answer C** is the right choice among those given here. Certain transmitters use a balanced modulator, but it will not match the antenna. A Standing Wave Ratio (SWR) Bridge can help you measure if you have a match, but it will not make the match. The Q multiplier is a receiver circuit intended to improve selectivity and sensitivity, not impedance matching.

G4A07 What condition can lead to permanent damage to a solid-state RF power amplifier?
A. Insufficient drive power
B. Low input SWR
C. Shorting the input signal to ground
D. Excessive drive power

If you do not have a sufficiently strong signal, you may not get the desired output, but you generally will not damage the amplifier, so this is incorrect. A low SWR is a good thing, so it will not damage the amplifier. Grounding the input will not be productive because you will wipe out the input signal, but you will not harm the RF amplifier. However, excessive drive power, as in **Answer D**, can damage the solid-state amplifier, so this is the right choice.

G4A08 What is the correct adjustment for the load or coupling control of a vacuum tube RF power amplifier?
A. Minimum SWR on the antenna
B. Minimum plate current without exceeding maximum allowable grid current
C. Highest plate voltage while minimizing grid current
D. Maximum power output without exceeding maximum allowable plate current

Yes, radio electronics applications still use vacuum tubes, especially for high-power signal amplification. Plate currents and grid currents are two critical parameters in vacuum tube circuit design. You should be able to guess that maximizing power output from a minimum circuit input might be something that radio designers are interested in, and **Answer D** is the right choice.

G4A09 Why is a time delay sometimes included in a transmitter keying circuit?
  A. To prevent stations from interfering with one another
  B. To allow the transmitter power regulators to charge properly
  C. To allow time for transmit-receive changeover operations to complete properly before RF output is allowed
  D. To allow time for a warning signal to be sent to other stations

Many transmitter/receiver systems need to set a relay to switch from "transmit" mode to "receive" mode, as in Figure 4.4. This relay switching takes a small amount of time to change connections. If the transmitter starts emitting RF too quickly, the emission can harm the receiver's electronics. Therefore, the circuit designers insert a slight delay to prevent the transmitter from signaling before the receiver is safely isolated, as in **Answer C**.

G4A10 What is the purpose of an electronic keyer?
  A. Automatic transmit/receive switching
  B. Automatic generation of strings of dots and dashes for CW operation
  C. VOX operation
  D. Computer interface for PSK and RTTY operation

A keyer is used with CW operations, so Answers C and D cannot be correct. It sounds like it could be part of the transmit/receive switching, as in Answer A, but that is incorrect in amateur radio practice. The electronic keyer assists in sending the actual Morse code symbols making **Answer B** the right choice.

G4A11 Which of the following is a use for the IF shift control on a receiver?
  A. To avoid interference from stations very close to the receive frequency
  B. To change frequency rapidly
  C. To permit listening on a different frequency from that on which you are transmitting
  D. To tune in stations that are slightly off frequency without changing your transmit frequency

Unless you have a HF rig that you have used frequently, all of the choices might look good. However, as **Answer A** states, the Intermediate Frequency (IF) shift circuit helps the operator avoid interference from a station near the receive frequency. The shift is a small amount, so it will not help change frequencies rapidly. Answer C has the definition of "split mode," so do not confuse that here. It is not a fine-tuning aid, as in Answer D.

G4A12 Which of the following is a common use for the dual-VFO feature on a transceiver?
A. To allow transmitting on two frequencies at once
B. To permit full duplex operation – that is, transmitting and receiving at the same time
C. To permit monitoring of two different frequencies
D. To facilitate computer interface

One generally does not transmit two frequencies simultaneously, so Answer A is incorrect. Answer D is technobabble. A dual VFO is an oscillator that allows the operator to monitor two different frequencies at the same time, so **Answer C** is correct. Amateur radio communications do not operate in full-duplex mode.

G4A13 What is one reason to use the attenuator function that is present on many HF transceivers?
A. To reduce signal overload due to strong incoming signals
B. To reduce the transmitter power when driving a linear amplifier
C. To reduce power consumption when operating from batteries
D. To slow down received CW signals for better copy

Believe it or not, sometimes an incoming signal can be too strong, and it will overload your transceiver's RF input. In these cases, it is helpful to have an attenuator built into the XCVR to reduce the signal strength, as in **Answer A**. It is not an output attenuator, so Answer B is incorrect. Answers C and D are just distractions.

G4A14 What is likely to happen if a transceiver's ALC system is not set properly when transmitting AFSK signals with the radio using single sideband mode?
A. ALC will invert the modulation of the AFSK mode
B. Improper action of ALC distorts the signal and can cause spurious emissions
C. When using digital modes, too much ALC activity can cause the transmitter to overheat
D. All of these choices are correct

Since digital modes pop up in several questions, you may suspect that they are popular. They are, and you should try at least one! However, one of the tricks that a good operator knows is that the transceiver's ALC should not activate when sending many of the digital modes; otherwise, the signal will not be as clean as it should be. **Answer B** is the right choice. Answers A and C are technically incorrect statements, making Answer D wrong.

G4A15 Which of the following can be a symptom of transmitted RF being picked up by an audio cable carrying AFSK data signals between a computer and a transceiver?
- A. The VOX circuit does not un-key the transmitter
- B. The transmitter signal is distorted
- C. Frequent connection timeouts
- D. All of these choices are correct

Each of the effects in Answers A, B, and C is a possible result of this problem, so **Answer D** is the best choice.

G4A16 How does a noise blanker work?
- A. By temporarily increasing received bandwidth
- B. By redirecting noise pulses into a filter capacitor
- C. By reducing receiver gain during a noise pulse
- D. By clipping noise peaks

The noise blanker reduces the receiver's gain during a pulse, which makes **Answer C** the correct choice. Increasing the bandwidth will only add more noise, so this is not a good choice. Filtering capacitors react to the whole input bandwidth, so redirection is impossible. Clipping the noise pulse will still permit extra noise into the receiver, which is not helpful.

G4A17 What happens as the noise reduction control level in a receiver is increased?
- A. Received signals may become distorted
- B. Received frequency may become unstable
- C. CW signals may become severely attenuated
- D. Received frequency may shift several kHz

Noise reduction is a form of filtering, and this filtering can start to distort the signals by removing components. This property makes **Answer A** the correct choice. The action will not change frequencies or shift the signal, so Answers B and D are incorrect. It does not attenuate the CW signals as in Answer C.

4.4   G4B - Testing

4.4.1   Overview

The *Testing* question group in Subelement G4 quizzes you on component testing principles for radio circuits. The *Testing* group covers topics such as
- Test and monitoring equipment
- Two-tone test

The test producer will select one of the 15 questions in this group for your exam.

4.4.2 Questions

G4B01 What item of test equipment contains horizontal and vertical channel amplifiers?
A. An ohmmeter
B. A signal generator
C. An ammeter
D. An oscilloscope

If you know about test equipment, then you will recognize the correct choice is the oscilloscope in **Answer D**. Both ohmmeters and ammeters are single-channel devices, so Answers A and C are incorrect. A signal generator produces signals, not displays signals, so Answer B is incorrect.

G4B02 Which of the following is an advantage of an oscilloscope versus a digital voltmeter?
A. An oscilloscope uses less power
B. Complex impedances can be easily measured
C. Input impedance is much lower
D. Complex waveforms can be measured

An oscilloscope measures and dispalys complex waveforms in the time domain, not impedances, so **Answer D** is the correct choice and Answer B is wrong. Answer A is incorrect, and in fact, the opposite is true. Both meters should have very high input impedances, so Answer C is wrong.

G4B03 Which of the following is the best instrument to use when checking the keying waveform of a CW transmitter?
A. An oscilloscope
B. A field strength meter
C. A sidetone monitor
D. A wavemeter

The oscilloscope of **Answer A** is the best option for checking the signal's waveform. A field-strength meter is for measuring radiated RF signals. A wavemeter can tell you if the signal is present, not the signal's quality, so it is also incorrect. The sidetone monitor in Answer C will generally let you hear your generated CW keying but not see the transmitted waveform.

G4B04 What signal source is connected to the vertical input of an oscilloscope when checking the RF envelope pattern of a transmitted signal?
A. The local oscillator of the transmitter
B. An external RF oscillator
C. The transmitter balanced mixer output
D. The attenuated RF output of the transmitter

If you wish to check the transmitted signal, you should check the RF final output, as in **Answer D**. You will need an attenuator to keep from overloading the 'scope. One does not make the desired measurement at the locations in Answers A and C because they are not at the final stage of the transmission electronics chain. They occur earlier, and "bad things" can still happen after them. Answer B is to distract you.

G4B05 Why is high input impedance desirable for a voltmeter?
   A. It improves the frequency response
   B. It decreases battery consumption in the meter
   C. It improves the resolution of the readings
   D. It decreases the loading on circuits being measured

High impedance means that the meter will not put a load on the circuit elements it measures. This property makes **Answer D** the right choice. Answers A, B, and C have nothing to do with the meter's input impedance, so these are incorrect.

G4B06 What is an advantage of a digital voltmeter as compared to an analog voltmeter?
   A. Better for measuring computer circuits
   B. Better for RF measurements
   C. Better precision for most uses
   D. Faster response

A quality digital voltmeter will give the user a result with higher precision than an analog voltmeter, so **Answer C** is the best choice here. The other options can be performance "ties" depending upon the quality of the respective meters and the exact nature of the measurement.

G4B07 What signals are used to conduct a two-tone test?
   A. Two audio signals of the same frequency shifted 90 degrees
   B. Two non-harmonically related audio signals
   C. Two swept frequency tones
   D. Two audio frequency range square wave signals of equal amplitude

The transmitter linearity test uses two non-harmonically related tones in the audio passband to drive the transmitter, so **Answer B** is the correct formulation of the test conditions. Answer A is incorrect because the tones are at the same frequency. Answer C is wrong because the tones need to be non-harmonically related, so any arbitrary tones generated by the sweep will not work consistently. Answer D is incorrect because the square waves produce many harmonically-related tones.

G4B08 Which of the following instruments may be used to monitor relative RF output when making antenna and transmitter adjustments?
  A. A field strength meter
  B. An antenna noise bridge
  C. A multimeter
  D. A Q meter

An antenna field strength meter will give you an indication of the strength of the RF signal coming from the antenna, so **Answer A** is correct. Operators use a noise bridge to measure the antenna matching, not the field strength. One uses a multimeter for measuring voltage, current, and resistance while a Q meter measures circuit resonance.

G4B09 Which of the following can be determined with a field strength meter?
  A. The radiation resistance of an antenna
  B. The radiation pattern of an antenna
  C. The presence and amount of phase distortion of a transmitter
  D. The presence and amount of amplitude distortion of a transmitter

The field strength meter measures the strength of the radiation coming off an antenna. By moving the meter around, you can trace out the antenna's pattern, making **Answer B** the right choice. It will not measure radiation resistance. It will not measure the distortions in the transmitter either.

G4B10 Which of the following can be determined with a directional wattmeter?
  A. Standing wave ratio
  B. Antenna front-to-back ratio
  C. RF interference
  D. Radio wave propagation

A directional wattmeter gives you a measurement of transmitted and reflected signals that you can use to find the antenna's SWR, so **Answer A** is the right choice. You can use the field strength meter to find the antenna's front-to-back ratio. A single piece of test equipment usually cannot make the measurements for Answers C and D.

G4B11 Which of the following must be connected to an antenna analyzer when it is being used for SWR measurements?
  A. Receiver
  B. Transmitter
  C. Antenna and feed line
  D. All of these choices are correct

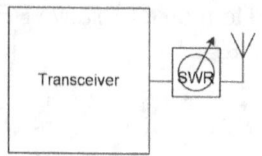

Figure   4.5:   SWR measurement configuration.

This is one of those obvious questions once you think about it. As Figure 4.5 shows, to measure the antenna's SWR, you need to have the antenna and feedline connected, so **Answer C** is the right choice. Answers A and B will not tell you about the antenna's SWR. Since Answers A and B are incorrect, Answer D must be wrong.

G4B12 What problem can occur when making measurements on an antenna system with an antenna analyzer?
- A. Permanent damage to the analyzer may occur if it is operated into a high SWR
- B. Strong signals from nearby transmitters can affect the accuracy of measurements
- C. The analyzer can be damaged if measurements outside the ham bands are attempted
- D. Connecting the analyzer to an antenna can cause it to absorb harmonics

If the manufacturer has appropriately constructed the analyzer, the measurement should not be affected by a high SWR. If one uses the analyzer over its manufacturer's recommended range, it will not matter if you use it outside the ham bands. Answer D is technobabble. Interference from nearby strong signals is a legitimate concern, so **Answer B** is the correct choice.

G4B13 What is a use for an antenna analyzer other than measuring the SWR of an antenna system?
- A. Measuring the front to back ratio of an antenna
- B. Measuring the turns ratio of a power transformer
- C. Determining the impedance of coaxial cable
- D. Determining the gain of a directional antenna

An antenna analyzer is basically an impedance measurement device, so determining the impedance of a coaxial cable is within its normal range of activities, and **Answer C** is the right choice. The other answers do not contain suggested measurements that an antenna analyzer might realistically perform.

G4B14 What is an instance in which the use of an instrument with analog readout may be preferred over an instrument with digital readout?
- A. When testing logic circuits
- B. When high precision is desired
- C. When measuring the frequency of an oscillator
- D. When adjusting tuned circuits

As we saw earlier, a digital readout is usually preferred when making a high-precision measurement, and we can eliminate Answer B. However, when tuning a circuit and adjusting it near the tuning point, an analog readout is preferred by many users because it gives them a better "feel" for how the process is going. Therefore, **Answer D** is the correct choice. Logic circuits are digital, so usually, people prefer a digital readout. When measuring the exact frequency, users often desire a numeric readout.

G4B15 What type of transmitter performance does a two-tone test analyze?
A. Linearity
B. Percentage of suppression of carrier and undesired sideband for SSB
C. Percentage of frequency modulation
D. Percentage of carrier phase shift

As we saw earlier, a two-tone test measures the linearity of the transmitter, which makes **Answer A** the right choice. It will not make any of the measurements given as the choices in Answers B, C, or D, so these are incorrect.

4.5 G4C - Interference

4.5.1 Overview

The *Interference* question group in Subelement G4 examines sources and remedies for interference in radio circuits. The *Interference* group covers topics such as
- Interference with consumer electronics
- grounding
- Digital Signal Processing

The test producer will select one of the 13 questions in this group for your exam.

4.5.2 Questions

G4C01 Which of the following might be useful in reducing RF interference to audio frequency devices?
A. Bypass inductor
B. Bypass capacitor
C. Forward-biased diode
D. Reverse-biased diode

We need to understand a bit of electrical theory to answer this question. At the relatively high frequencies of radio signals, inductors tend to have higher impedance than do capacitors. Suppose one connects a capacitor between the input and ground. The radio signal causing the interference will see a low-impedance path to the ground (shunted to ground) for removal. This result is the desired action, and the bypass capacitor of **Answer B** is the right choice. The inductor placed as a bypass circuit element will have high impedance and not work. Diodes are ineffective here because the RF signal has both positive and negative amplitudes, so the ones in Answers C and D will only conduct in one direction and not entirely remove the signal.

G4C02 Which of the following could be a cause of interference covering a wide range of frequencies?
  A. Not using a balun or line isolator to feed balanced antennas
  B. Lack of rectification of the transmitter's signal in power conductors
  C. Arcing at a poor electrical connection
  D. Using a balun to feed an unbalanced antenna

Using or not using a balun may make for a high SWR but will not cause interference, so Answers A and D are not good choices. Having rectification in power lines is a problem, so the lack of it is good, making Answer B incorrect. Arcing will produce a wide-bandwidth interference signal and be a safety problem, so **Answer C** is the correct choice.

G4C03 What sound is heard from an audio device or telephone if there is interference from a nearby single sideband phone transmitter?
  A. A steady hum whenever the transmitter is on the air
  B. On-and-off humming or clicking
  C. Distorted speech
  D. Clearly audible speech

The correct answer here should be easy to spot. If one is transmitting phone using Single Sideband (SSB), then the original signal is a phone transmission, and the interference should be some form of a speech signal. Because the device is not a proper SSB receiver, then the speech signal will be distorted, as in **Answer C**. Answer D is the opposite. The steady hum of Answer A is often associated with a grounding problem. The clicks of Answer B are from CW signals.

G4C04 What is the effect on an audio device when there is interference from a nearby CW transmitter?
  A. On-and-off humming or clicking
  B. A CW signal at a nearly pure audio frequency
  C. A chirpy CW signal
  D. Severely distorted audio

This question is similar to the previous one. Severely distorted audio is not specific enough, so this is not the best choice for the answer. Because the device is not a CW receiver, we would not expect the output to be clean, so Answer B is not a good choice. You need to remember that the signal will sound like an on-off hum or click, as in **Answer A**. This situation does not produce the chirp of Answer C.

G4C05 What might be the problem if you receive an RF burn when touching your equipment while transmitting on an HF band, assuming the equipment is connected to a ground rod?
  A. Flat braid rather than round wire has been used for the ground wire
  B. Insulated wire has been used for the ground wire
  C. The ground rod is resonant
  D. The ground wire has high impedance on that frequency

The question is setting up the case where the ground system does not conduct the stray RF energy to the ground rod. A flat braid is frequently used, which makes for a good ground connection, so Answer A is an incorrect statement. The insulation will not prevent RF energy from being conducted, making Answer B wrong as well. If the ground rod is a resonant length, that will not keep it from conducting the RF energy to the ground. However, if the ground wire has a high impedance, it will not conduct the RF energy to the ground rod effectively, so **Answer D** is the correct choice.

G4C06 What effect can be caused by a resonant ground connection?
  A. Overheating of ground straps
  B. Corrosion of the ground rod
  C. High RF voltages on the enclosures of station equipment
  D. A ground loop

A resonant ground connection acts like an antenna and picks up stray RF energy. The stray energy will produce high RF voltages on the equipment, as in **Answer C**. None of the other choices represent a true statement under these conditions.

G4C07 Why should soldered joints not be used with the wires that connect the base of a tower to a system of ground rods?
  A. The resistance of solder is too high
  B. Solder flux will prevent a low conductivity connection
  C. Solder has too high a dielectric constant to provide adequate lightning protection
  D. A soldered joint will likely be destroyed by the heat of a lightning strike

Solder is an alloy with a relatively low melting point. The heat from a lightning strike could cause melting making the joint fail. This result makes **Answer D** the right choice. The other options are incorrect electrically.

G4C08 Which of the following would reduce RF interference caused by common-mode current on an audio cable?
A. Placing a ferrite choke around the cable
B. Adding series capacitors to the conductors
C. Adding shunt inductors to the conductors
D. Adding an additional insulating jacket to the cable

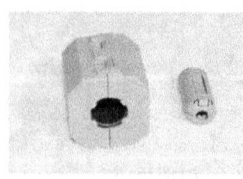

A ferrite choke, like the one shown in Figure 4.6, is used to reduce RF interference on cables. This choke makes **Answer A** the correct choice. The other answers are not electrically correct choices to remove RF energy on a cable.

Figure 4.6: Example ferrite chokes.

G4C09 How can a ground loop be avoided?
A. Connect all ground conductors in series
B. Connect the AC neutral conductor to the ground wire
C. Avoid using lock washers and star washers when making ground connections
D. Connect all ground conductors to a single point

The best remedy is to have a single ground reference point, as in **Answer D**. Answer A may sound like the same thing, but it is not electrically. Answer C is a silly distraction. Connecting the Alternating Current (AC) neutral to the ground is typical, but this does not help with battery-powered or Direct Current (DC) equipment.

G4C10 What could be a symptom of a ground loop somewhere in your station?
A. You receive reports of "hum" on your station's transmitted signal
B. The SWR reading for one or more antennas is suddenly very high
C. An item of station equipment starts to draw excessive amounts of current
D. You receive reports of harmonic interference from your station

A ground loop symptom is a hum on your transmitted signal, so **Answer A** is the correct choice. A ground loop does not increase the SWR. It will not cause an increase in power consumption either. Be careful with Answer D because it sounds close, but it is not as specific as Answer A.

G4C11 What technique helps to minimize RF "hot spots" in an amateur station?
  A. Building all equipment in a metal enclosure
  B. Using surge suppressor power outlets
  C. Bonding all equipment enclosures together
  D. Low-pass filters on all feed lines

RF hot spots occur when the ground connections do not provide complete electrical return paths. Bonding enclosures together, as in **Answer C**, is a way to avoid hot spots. Having a metal enclosure by itself will not prevent hot spots. Surge suppressors prevent stray signals from the power line and not the RF signals. The LPFs are for reducing any harmonic signals.

G4C12 Which of the following is an advantage of a receiver DSP IF filter as compared to an analog filter?
  A. A wide range of filter bandwidths and shapes can be created
  B. Fewer digital components are required
  C. Mixing products are greatly reduced
  D. The DSP filter is much more effective at VHF frequencies

The best thing about Digital Signal Processor (DSP) filters is that you can use them to create a variety of filters characteristics, as in **Answer A**. Each of the statements in Answers B, C, and D is incorrect.

G4C13 Why must the metal enclosure of every item of station equipment be grounded?
  A. It prevents a blown fuse in the event of an internal short circuit
  B. It prevents signal overload
  C. It ensures that the neutral wire is grounded
  D. It ensures that hazardous voltages cannot appear on the chassis

Proper grounding will prevent hazardous voltages from appearing on the chassis, making **Answer D** the right choice. It will not prevent blowing fuses due to shorts, which would be a bad feature. It also will not prevent signal overloads or ensure the neutral wire is connected.

4.6   G4D - Transceiver Operations

4.6.1   Overview

The *Transceiver Operations* question group in Subelement G4 covers signal processing operations in transceivers. The *Transceiver Operations* group covers topics such as
  • Speech processors
  • S meters

- Sideband operation near band edges

The test producer will select one of the 11 questions in this group for your exam.

4.6.2   Questions

G4D01 What is the purpose of a speech processor as used in a modern transceiver?
- A. Increase the intelligibility of transmitted phone signals during poor conditions
- B. Increase transmitter bass response for more natural sounding SSB signals
- C. Prevent distortion of voice signals
- D. Decrease high-frequency voice output to prevent out of band operation

Generally, the high frequencies in a person's voice will not cause out-of-band transmissions, so Answer D is incorrect. The speech processor is not to correct voice distortions, although Answer C sounds like a reason for a speech processor. It also is not for increasing bass response. Improving intelligibility, as in **Answer A**, is the right reason.

G4D02 Which of the following describes how a speech processor affects a transmitted single sideband phone signal?
- A. It increases peak power
- B. It increases average power
- C. It reduces harmonic distortion
- D. It reduces intermodulation distortion

The correct technical description of how a speech processor affects the transmitted signal is that it increases the average signal power, so **Answer B** is the right choice. It does not enhance large-amplitude signals, so it does not increase peak power, as in Answer A. Answers C and D would be nice to have, but they are not things a speech processor can do, so they are incorrect choices.

G4D03 Which of the following can be the result of an incorrectly adjusted speech processor?
- A. Distorted speech
- B. Splatter
- C. Excessive background pickup
- D. All of these choices are correct

Each effect in Answers A, B, and C is a possible result of having a poorly adjusted speech processor, so the best choice is **Answer D**.

G4D04 What does an S meter measure?
  A. Conductance
  B. Impedance
  C. Received signal strength
  D. Transmitter power output

You will find an S meter on your receiver, and you can use it to measure received signal strength, so **Answer C** is the right choice. Various multimeters measure the components in Answers A and B, so they are incorrect here. A different meter on your transceiver will typically estimate transmitted power.

G4D05 How does a signal that reads 20 dB over S9 compare to one that reads S9 on a receiver, assuming a properly calibrated S meter?
  A. It is 10 times less powerful
  B. It is 20 times less powerful
  C. It is 20 times more powerful
  D. It is 100 times more powerful

The dB scale is logarithmic, and a 20-dB increase is a factor of 100 increase in power, so **Answer D** is the correct choice. The other options are to see if you know what a dB means.

G4D06 Where is an S meter found?
  A. In a receiver
  B. In an SWR bridge
  C. In a transmitter
  D. In a conductance bridge

As we saw above, we find the S meter on a receiver, so **Answer A** is correct. The other choices are asking if you know your equipment.

G4D07 How much must the power output of a transmitter be raised to change the S meter reading on a distant receiver from S8 to S9?
  A. Approximately 1.5 times
  B. Approximately 2 times
  C. Approximately 4 times
  D. Approximately 8 times

The S meter's scale is not an exact measurement. The manufacturer usually designs it so that an increase of 1 S unit corresponds to a factor of 4 change in power. The correct choice then is **Answer C**.

G4D08 What frequency range is occupied by a 3 kHz LSB signal when the displayed carrier frequency is set to 7.178 MHz?
  A. 7.178 to 7.181 MHz
  B. 7.178 to 7.184 MHz
  C. 7.175 to 7.178 MHz
  D. 7.1765 to 7.1795 MHz

Because we are using Lower Side Band (LSB), the signal lies on the low side of the carrier. The lower edge of the signal will start at 7.178 MHz − 3 kHz = 7.178 MHz − 0.003 MHz = 7.175 MHz. The upper edge of the signal will stop at the carrier of 7.178 MHz. **Answer C** is the correct range. Be careful with Answer A because that is the range for Upper Side Band (USB).

G4D09 What frequency range is occupied by a 3 kHz USB signal with the displayed carrier frequency set to 14.347 MHz?
  A. 14.347 to 14.647 MHz
  B. 14.347 to 14.350 MHz
  C. 14.344 to 14.347 MHz
  D. 14.3455 to 14.3485 MHz

Now we switch to USB. The lower edge will be at the carrier frequency of 14.347 MHz. The upper edge will be at 14.347 MHz + 0.003 MHz = 14.350 MHz. **Answer B** has the correct range. Do you agree that Answer C has the range for LSB?

G4D10 How close to the lower edge of the phone segment should your displayed carrier frequency be when using 3 kHz wide LSB?
  A. At least 3 kHz above the edge of the segment
  B. At least 3 kHz below the edge of the segment
  C. At least 1 kHz below the edge of the segment
  D. At least 1 kHz above the edge of the segment

Using the previous questions as a guide, you should be able to pick out the correct choice. With LSB, the carrier needs to be at least 3 kHz above the lower band edge, as in **Answer A**. The other choices will have your signal going past the band edge.

G4D11 How close to the upper edge of the phone segment should your displayed carrier frequency be when using 3 kHz wide USB?
  A. At least 3 kHz above the edge of the band
  B. At least 3 kHz below the edge of the band
  C. At least 1 kHz above the edge of the segment
  D. At least 1 kHz below the edge of the segment

Using the reasoning we have developed over the past few questions, you should say that the carrier should be no closer than 3 kHz below the upper band edge. **Answer B** is the correct choice, and the others would cause you to go beyond the band edge.

4.7 G4E - Remotely Operating

4.7.1 Overview

The *Remotely Operating* question group in Subelement G4 looks at mobile radio operations. The *Remotely Operating* group covers topics such as
- HF mobile radio installations
- Alternative energy source operation

The test producer will select one of the 11 questions in this group for your exam.

4.7.2 Questions

G4E01 What is the purpose of a capacitance hat on a mobile antenna?
- A. To increase the power handling capacity of a whip antenna
- B. To allow automatic band changing
- C. To electrically lengthen a physically short antenna
- D. To allow remote tuning

Because mobile antennas tend to be electrically short for the HF bands, designers use a capacitance hat to electrically lengthen an antenna to make it resonant on the desired part of the frequency band. **Answer C** is the correct choice. The hat will not make the antenna handle more power, nor does it permit band changing or remote tuning.

G4E02 What is the purpose of a corona ball on a HF mobile antenna?
- A. To narrow the operating bandwidth of the antenna
- B. To increase the "Q" of the antenna
- C. To reduce the chance of damage if the antenna should strike an object
- D. To reduce RF voltage discharge from the tip of the antenna while transmitting

Engineers often refer to a high-voltage discharge as "corona." We find the highest voltages on an antenna at the tip. A corona ball reduces high voltage discharge from the tip of the antenna while transmitting, as in **Answer D**. The other choices are to distract you.

G4E03 Which of the following direct, fused power connections would be the best for a 100 watt HF mobile installation?
- A. To the battery using heavy-gauge wire
- B. To the alternator or generator using heavy-gauge wire
- C. To the battery using resistor wire
- D. To the alternator or generator using resistor wire

A 100-W HF mobile rig will draw a reasonable amount of current and will need a wiring harness capable of handling the current. The resistor wire found in Answers C and D will be light-duty and not recommended for high current applications. Answer B is not good because the rig requires a DC input, and the alternator or generator produces an AC output. The heavy gauge wire connected to the battery in **Answer A** is the correct choice.

G4E04 Why is it best NOT to draw the DC power for a 100 watt HF transceiver from a vehicle's auxiliary power socket?
- A. The socket is not wired with an RF-shielded power cable
- B. The socket's wiring may be inadequate for the current drawn by the transceiver
- C. The DC polarity of the socket is reversed from the polarity of modern HF transceivers
- D. Drawing more than 50 watts from this socket could cause the engine to overheat

As we saw in the previous question, the 100-W rig will draw a relatively large current. Of the statements given, the best reason is **Answer B** because the car's manufacturer does not rate its wiring for high currents. Answers C and D are technically untrue. The RF shielding in Answer A is not necessary.

G4E05 Which of the following most limits an HF mobile installation?
- A. "Picket fencing" signal variation
- B. The wire gauge of the DC power line to the transceiver
- C. Efficiency of the electrically short antenna
- D. FCC rules limiting mobile output power on the 75-meter band

"Picket fencing" describes audio cut-out near the capture threshold of a receiver and is not a general characteristic across the entire band, so this is not a good choice. The wire gauge is not relevant to effective transmission as long as it is safe for the required current. Answer D does not match the Part 97 regulations. The antenna will be electrically short, and it will limit operations on the HF bands. Therefore, **Answer C** is the best choice. This choice makes sense because a full-size HF antenna is a significant fraction of the operating wavelength and will be considerably longer than the vehicle itself.

G4E06 What is one disadvantage of using a shortened mobile antenna as opposed to a full-size antenna?
- A. Short antennas are more likely to cause distortion of transmitted signals
- B. Short antennas can only receive circularly polarized signals
- C. Operating bandwidth may be very limited
- D. Harmonic radiation may increase

Shortened antennas are not resonant across the band. The designers add elements such as traps and hats to improve operations, but they are generally only suitable for a small band region. Therefore, operations are limited, as in **Answer C**. Each of the statements in Answers A, B, and D is technically incorrect.

G4E07 Which of the following may cause receive interference in a radio installed in a vehicle?
- A. The battery charging system
- B. The fuel delivery system
- C. The vehicle control computer
- D. All of these choices are correct

With all of the electronics in modern vehicles, each of the components in Answers A, B, and C can cause interference if they are operating incorrectly. This situation makes **Answer D** the right choice.

G4E08 What is the name of the process by which sunlight is changed directly into electricity?
- A. Photovoltaic conversion
- B. Photon emission
- C. Photosynthesis
- D. Photon decomposition

Photon decomposition is a silly distraction. Photosynthesis is valid for plants and is also silly in this context. Photon emission is the opposite, where a device converts electricity to light. Photovoltaic conversion, as in **Answer A**, is the right choice for this question.

G4E09 What is the approximate open-circuit voltage from a fully illuminated silicon photovoltaic cell?
- A. 0.02 VDC
- B. 0.5 VDC
- C. 0.2 VDC
- D. 1.38 VDC

You need to remember that the correct choice is 0.5 V, as in **Answer B**. The others are not typical for modern silicon photocells.

G4E10 What is the reason that a series diode is connected between a solar panel and a storage battery that is being charged by the panel?
  A. The diode serves to regulate the charging voltage to prevent overcharge
  B. The diode prevents self-discharge of the battery through the panel during times of low or no illumination
  C. The diode limits the current flowing from the panel to a safe value
  D. The diode greatly increases the efficiency during times of high illumination

When the sunlight level is low, the solar cell might have a lower voltage across it than the battery. The current could attempt to flow back across the solar cell in this situation. Designers use a diode to prevent this back current flow, and **Answer B** is the right choice. Diodes do not regulate voltage, so Answer A is incorrect. The diode does not limit the forward current flow, so Answer C is also wrong. Because the diode has a slight energy loss associated with it, it cannot increase efficiency, as in Answer D.

G4E11 Which of the following is a disadvantage of using wind as the primary source of power for an emergency station?
  A. The conversion efficiency from mechanical energy to electrical energy is less than 2 percent
  B. The voltage and current ratings of such systems are not compatible with amateur equipment
  C. A large energy storage system is needed to supply power when the wind is not blowing
  D. All of these choices are correct

A 2% conversion efficiency may be true in a given configuration but not as big of a disadvantage as the need for a power supply when the wind is not blowing. The voltage rating of the system is not necessarily incompatible. Since Answers A and B are incorrect, Answer D is also wrong. **Answer C** is the only option that is true all of the time and, therefore, is the best choice for this question.

Chapter 5

G5 — ELECTRICAL PRINCIPLES

5.1 Introduction

In the Technician Class study guide, we introduced Ohm's Law for resistors in Direct Current (DC) circuits and how to compute the power absorbed by the resistors. We also learned about the decibel power notation. For the General Class license study, we will expand that to look at capacitors and inductors in Alternating Current (AC) circuits as well. We will also look at basic circuit concepts with series and parallel circuits. The *Electrical Principles* subelement has the following question groups:

A. Reactive Elements
B. Power
C. Series and Parallel Circuits

Subelement 5 will generate three questions on the General Class examination.

5.2 Radio Engineering Concepts

Resistors, Capacitors, and Inductors From the Technician Class study guide, you should remember the relationship between resistors and Ohm's Law. As a refresher, we saw the three versions of Ohm's Law based on if we needed to solve for voltage, current, or resistance by using one of the following equations:

Voltage Form — *Voltage = Resistance × Current*
Current Form — *Current = Voltage ÷ Resistance*
Resistance Form — *Resistance = Voltage ÷ Current*

In that earlier study guide, we also saw that *capacitors* are components that store electrical energy in an electric field. In contrast, *inductors* store electrical energy in a magnetic field. What we need to learn here is that both of these

Table 5.1: Impedances for Perfect Electrical Components.

	Impedance		
Component	Resistance	Reactance	Unit
Resistor	R	0	Ω
Capacitor	0	$1/(2\pi f C)$	Ω
Inductor	0	$2\pi f L$	Ω

components are frequency-dependent and behave differently with signals in the High Frequency (HF) range than they do in the Very High Frequency (VHF) range.

With resistors, we measure the resistance in the unit of ohm (Ω). At first, when we measured that resistance, we used a DC source. Because capacitors and inductors behave differently with different frequencies, we will start thinking in terms of AC circuits as well. From circuit theory, we model any current as being composed of a DC component and an AC component. To match this, we model general circuit elements with a resistive part, R, for the DC component and a reactive part, X, for the AC component. Now, instead of resistance, we will use the more general term *impedance* in this general environment. Table 5.1 lists the resistive and reactive components for perfect resistors, capacitors, and inductors. Notice that both the resistive and reactive components have the units of ohms. Ideal resistors contain no reactive part, while ideal capacitors and inductors contain only reactive parts. Real devices will often have a slight component mixture, such as a resistor with a small inductive component.

Power and dBs   Referring back to the Technician Class guide, we saw that the power in watts, P, dissipated by a resistor is given in terms of the current through the resistor, I, and the voltage across the resistor, V, by

$$P = V \times I$$
$$P = V^2 \div R$$
$$P = I^2 \times R$$

This definition extends to any electrical circuit element if we compute it correctly. With reactive elements and AC circuits, these equations hold for any specific instant in time. We need to make a slight modification to these equations to have a more representative measurement at all times. To adjust for this time variability, engineers use the Root Mean Square (RMS) values for the current or the voltage. Mathematically, this gives the DC equivalent for the AC values. For a sinusoidal source, the peak voltage value is called the Peak Envelope Voltage (PEV). The sinusoidal RMS voltage, V_{RMS}, is computed as $V_{RMS} = PEV/\sqrt{2} = 0.707\,PEV$. Figure 5.1 illustrates peak and RMS levels for single and two-toned sinusoid waveforms to show this difference. In both cases,

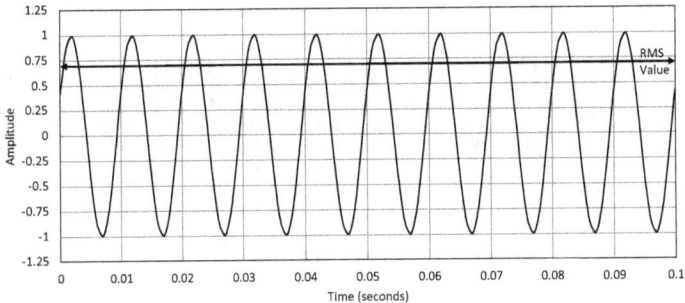

(a) RMS signal level for a single-tone sinusoid.

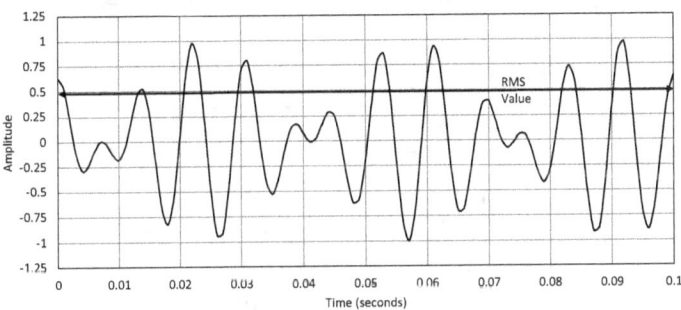

(b) RMS signal level for a two-tone sinusoid.

Figure 5.1: RMS and peak levels for single and two-tone sinusoidal waveforms having peaks of 1 V.

the PEV is at 1 V.

With this information, we can now compute power for AC circuit elements. If we call the general impedance Z for the circuit element, we compute the RMS power by one of the three equivalent equations:

$$P_{RMS} = V_{RMS} \times I_{RMS}$$
$$P_{RMS} = V_{RMS}^2 \div Z$$
$$P_{RMS} = I_{RMS}^2 \times Z$$

Here, Z can be purely resistive, purely reactive, or a combination of both.

Because we have time-varying signals, Part 97 makes a special power definition of the Peak Envelope Power (PEP) for the output of transmitters. Part 97 defines PEP as the "average power supplied to the antenna transmission line by a transmitter during one RF cycle at the crest of the modulation envelope taken under normal operating conditions." Figure 5.2 shows the envelope for an

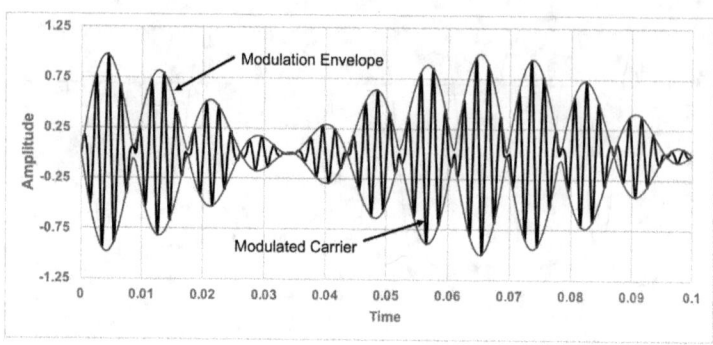

Figure 5.2: Modulation envelope for an AM signal.

Table 5.2: Common Decibel Values Relative to the Reference Power.

Gain Factor	dB	Loss Factor	dB
2	3	½	-3
4	6	¼	-6
10	10	0.1	-10
100	20	0.01	-20
1000	30	0.001	-30
1000000	60	0.000001	-60

Amplitude Modulation (AM) modulated signal. The envelope is the "wrapper" that includes the modulated carrier. Notice that the envelope's amplitude level changes with time, as does the output power. For example, the single-cycle RMS power around 0.065 s is much higher than the single-cycle RMS power around 0.035 s. The envelope's peak average power value represents a "worst case" estimate for the transmission power. For constant-envelope signals like Frequency Modulation (FM) modulation or an unmodulated carrier, the RMS power is the same as the PEP.

Another refresher from the Technician Class study is the use of the decibel (dB) representation of a power ratio. The decibel power ratio is computed using $Power(dB) = 10 \log (DevicePower \div ReferencePower)$. Remember, if we compute power relative to 1 W the result is $P(dB) = 10 \log [(V \times I) / 1\,W]$. We often indicate this as dBW. Similarly, if we referenced to 1 mW it would be dBm. Table 5.2 lists common dB values for increasing or decreasing by common factors, such as 2 or 100, relative to the reference power.

Series and Parallel Circuits   There are two important circuit configurations that you need to learn: series and parallel. Series and parallel elements are defined as

Series — circuit elements in series have the same current flowing through them

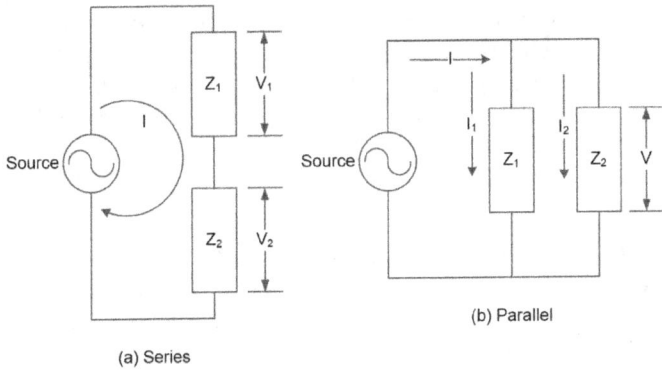

Figure 5.3: Series and parallel circuit configurations.

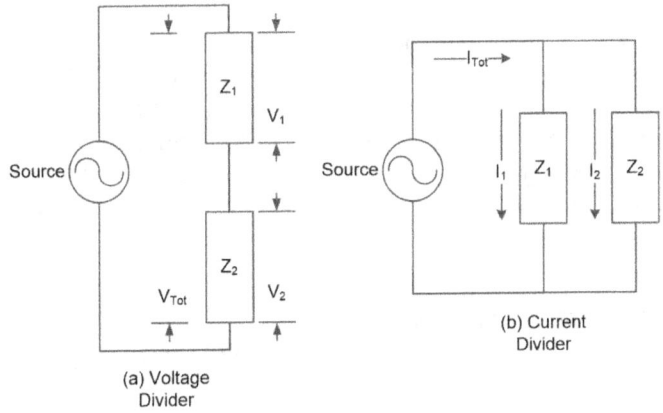

Figure 5.4: Voltage and current dividers.

Parallel — circuit elements in parallel have the same voltage across them
Figure 5.3 illustrates the series and parallel configuration. The current, I, that
flows through the series elements is sometimes called a *mesh current*. The total
current in the parallel circuit is composed of two *branch currents* (I_1 and I_2) that
flow through the parallel circuit elements.

Now we need to apply resistors, R, capacitors, C, and inductors, L, to these
basic building blocks. For series circuits
Total Series Resistance $R_{Total} = R_1 + R_2$
Total Series Inductance $L_{Total} = L_1 + L_2$
Total Series Capacitance $C_{Total} = 1/ \left(1/C_1 + 1/C_2 \right)$
For parallel circuits
Total Parallel Resistance $R_{Total} = 1/ \left(1/R_1 + 1/R_2 \right)$
Total Parallel Inductance $L_{Total} = 1/ \left(1/L_1 + 1/L_2 \right)$
Total Parallel Capacitance $C_{Total} = C_1 + C_2$

We can use these series and parallel configurations to construct two basic building blocks for electronics. The *voltage divider*, which Figure 5.4 shows in part (a), is based on the series circuit. We use different component values to divide the source voltage into specific proportions. The *current divider*, which Figure 5.4 shows in part (b), is based on the parallel circuit. We use different component values to divide the source current into specific proportions.

Transformers   A transformer is a special form of an inductive circuit element. The transformer is made from two closely-placed inductors so that their individual magnetic fields interact. This property is called *mutual inductance*. Manufacturers make each inductor from a coil of wire. One inductor is the "primary," and the other is the "secondary." Typically, the inductor coils have differing numbers of wire turns which cause either an increase (*step up*) or a decrease (*step down*) in voltage from the primary to the secondary. The relative number of wire turns is usually expressed as a turns ratio, N, e.g., 4:1, with N wire turns in the primary for every one wire turn in the secondary. The output voltage on the secondary, V_S is computed from N and the input voltage on the primary, V_P, by $V_S = V_P/N$. If we wish to use the transformer to match impedances, we use the square root of the turns ratio, not the ratio itself. We will apply this in a later question.

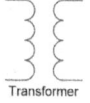

Transformer

Figure 5.5:
Circuit
symbol
for a trans-
former.

5.3   G5A - Reactive Elements

5.3.1   Overview

The *Reactive Elements* question group in Subelement G5 tests you on the electrical properties of capacitors and inductors. The *Reactive Elements* group covers topics such as
 • Capacitance and inductance
 • Reactance, impedance, and impedance matching
The test producer will select one of the 11 questions in this group for your exam.

5.3.2   Questions

G5A01 What is impedance?
 A. The electric charge stored by a capacitor
 B. The inverse of resistance
 C. The opposition to the flow of current in an AC circuit
 D. The force of repulsion between two similar electric fields

The Impedance (Z) is a generalized form of the Resistance (R), and it is any opposition to the flow of AC current. The correct choice is **Answer C**. The other options are electrically incorrect.

G5A02 What is reactance?
A. Opposition to the flow of direct current caused by resistance
B. Opposition to the flow of alternating current caused by capacitance or inductance
C. A property of ideal resistors in AC circuits
D. A large spark produced at switch contacts when an inductor is de-energized

The Reactance (X) is the frequency-dependent component of the impedance for inductors and capacitors. The reactance is any opposition to the flow of AC current. The correct choice is **Answer B**. Reactance does not apply to resistors, so do not choose Answers A or C. Answer D is a silly distraction.

G5A03 Which of the following causes opposition to the flow of alternating current in an inductor?
A. Conductance
B. Reluctance
C. Admittance
D. Reactance

From the previous questions, you should be able to spot the reactance in **Answer D** as the right choice. Reluctance is the opposition to magnetic flux for magnetic currents, which is similar to impedance. Conductance (G) is the reciprocal of the resistance, and Admittance (Y) is the reciprocal of the impedance.

G5A04 Which of the following causes opposition to the flow of alternating current in a capacitor?
A. Conductance
B. Reluctance
C. Reactance
D. Admittance

Here, you should be able to spot the reactance in **Answer C** as the right choice. Again, reluctance is the opposition to magnetic flux. Conductance and admittance are actual electrical quantities, but they are not the best choice here.

G5A05 How does an inductor react to AC?
A. As the frequency of the applied AC increases, the reactance decreases
B. As the amplitude of the applied AC increases, the reactance increases
C. As the amplitude of the applied AC increases, the reactance decreases
D. As the frequency of the applied AC increases, the reactance increases

The signal's frequency, not its amplitude, determines the properties of inductors and capacitors, making Answers B and C incorrect. Inductors and capacitors respond to the signal frequency in opposite ways. An inductor, L, has a higher opposition to current flow (higher impedance X_L) as the frequency of the signal increases ($X_L = 2\pi f L$). This property is why an inductor is sometimes called a choke, and the correct choice is **Answer D**. Answer A is for a capacitor, so be careful.

G5A06 How does a capacitor react to AC?
A. As the frequency of the applied AC increases, the reactance decreases
B. As the frequency of the applied AC increases, the reactance increases
C. As the amplitude of the applied AC increases, the reactance increases
D. As the amplitude of the applied AC increases, the reactance decreases

Now we examine the capacitor. We can eliminate Answers C and D from consideration since they deal with amplitude. A capacitor, C, has a higher opposition to current flow (higher impedance X_C) as the frequency of the signal decreases ($X_C = 1/(2\pi f C)$). This property is why designers sometimes use a capacitor to block DC, and the correct choice is **Answer A**. Answer B is for an inductor, so be careful not to choose it here.

G5A07 What happens when the impedance of an electrical load is equal to the output impedance of a power source, assuming both impedances are resistive?
A. The source delivers minimum power to the load
B. The electrical load is shorted
C. No current can flow through the circuit
D. The source can deliver maximum power to the load

Electrically, this condition produces a maximum power transfer between the source and the load, which is good. You should be able to spot **Answer D** as the right choice to answer this question. Be careful with Answer A because that is the opposite condition. Answers B and C are electrically incorrect.

G5A08 What is one reason to use an impedance matching transformer?
A. To minimize transmitter power output
B. To maximize the transfer of power
C. To reduce power supply ripple
D. To minimize radiation resistance

This question is asking about the maximum power principle in circuit analysis. This principle makes **Answer B** the right choice for this question. Answer A is the opposite of the desired effect, while Answers C and D are electrically wrong.

G5A09 What unit is used to measure reactance?
  A. Farad
  B. Ohm
  C. Ampere
  D. Siemens

We measure the Reactance (X) of a device in ohm (Ω), so the correct choice is **Answer B**. The ampere (A) is for current, the farad (F) is to measure capacitance, and Siemens (S) are for conductance, the reciprocal of resistance.

G5A10 Which of the following devices can be used for impedance matching at radio frequencies?
  A. A transformer
  B. A Pi-network
  C. A length of transmission line
  D. All these choices are correct

We can use each method listed in Answers A, B, and C for impedance matching. The multiple choices make **Answer D** the best option to answer the question.

G5A11 Which of the following describes one method of impedance matching between two AC circuits?
  A. Insert an LC network between the two circuits
  B. Reduce the power output of the first circuit
  C. Increase the power output of the first circuit
  D. Insert a circulator between the two circuits

Inserting the inductor-capacitor (LC) network between the circuits, as in **Answer A**, is a typical technique for impedance matching. Changing power levels will not affect the impedance difference between the circuits. A circulator is a device used to control the flow of Radio Frequency (RF) signals.

5.4 G5B - Power

5.4.1 Overview

The *Power* question group in Subelement G5 quizzes you on power concepts in radio circuits. The *Power* group covers topics such as
  • The Decibel
  • Current and voltage dividers
  • Electrical power calculations
  • Sine wave root-mean-square (RMS) values
  • PEP calculations
The test producer will select one of the 14 questions in this group for your exam.

5.4.2   Questions

G5B01 What dB change represents a factor of two increase or decrease in power?
  A. Approximately 2 dB
  B. Approximately 3 dB
  C. Approximately 6 dB
  D. Approximately 12 dB

You need to remember that a 2-times change in power is a 3-dB change ($10\log(2) = 3$ dB), so the correct choice is **Answer B**. The others do not compute properly. Be careful with 6 dB because it is a factor of 4 or 1 S unit.

G5B02 How does the total current relate to the individual currents in each branch of a purely resistive parallel circuit?
  A. It equals the average of each branch current
  B. It decreases as more parallel branches are added to the circuit
  C. It equals the sum of the currents through each branch
  D. It is the sum of the reciprocal of each individual voltage drop

When a circuit has multiple elements in parallel, as in Figure 5.4, the current splits among the branch elements, and the sum of the branch currents equals the total current. **Answer C** is the correct choice. Answer A deals with the average current, not the total current. Answers B and D are electrically wrong.

G5B03 How many watts of electrical power are used if 400 VDC is supplied to an 800 ohm load?
  A. 0.5 watts
  B. 200 watts
  C. 400 watts
  D. 3200 watts

Power is voltage, V, times the current, I, ($P = voltage \times current$). Using Ohm's Law, the power is $P = 400 \times 400 \div 800 = 200$ W, as in **Answer B**. Answers A, C, and D represent math mistakes. Answer A is $V \div R$, and not $V^2 \div R$, as required. Answer D is the product $V \times R$, and not $V \times I$ or $I^2 \times R$, as is required.

G5B04 How many watts of electrical power are used by a 12 VDC light bulb that draws 0.2 amperes?
  A. 2.4 watts
  B. 24 watts
  C. 6 watts
  D. 60 watts

If we multiply 12 V volts by 0.2 A, we get a result of 2.4 W or **Answer A**. Be careful with Answer B because it has the right digits but the wrong decimal place. Answers C and D are more math mistakes. Answer D is $V \div I$, and not $V \times I$, as needed for the right answer.

G5B05 How many watts are dissipated when a current of 7.0 milliamperes flows through 1250 ohm resistance?
A. Approximately 61 milliwatts
B. Approximately 61 watts
C. Approximately 11 milliwatts
D. Approximately 11 watts

Power is the square of the current times the resistance ($P = I^2 \times R$). We can compute that 7 mA squared times 1250 Ω will give 61 mW of power making **Answer A** the right choice. Answer B is missing the "milli."

G5B06 What is the output PEP from a transmitter if an oscilloscope measures 200 volts peak-to-peak across a 50 ohm dummy load connected to the transmitter output?
A. 1.4 watts
B. 100 watts
C. 353.5 watts
D. 400 watts

The PEP is the largest average power over a transmitted carrier cycle so we start with the peak-to-peak voltage. The peak envelope voltage is ½ the peak-to-peak voltage or 100 V. The PEP is $PEP = PEV^2 \div 2R_L$ or $100^2 \div (2 \times 50)$ or 100 W. The correct choice is **Answer B**. The other answers have math errors.

G5B07 What value of an AC signal produces the same power dissipation in a resistor as a DC voltage of the same value?
A. The peak-to-peak value
B. The peak value
C. The RMS value
D. The reciprocal of the RMS value

The RMS value provides the DC equivalent for an AC signal, so **Answer C** is the correct choice. The others are good measurements, but incorrect here.

G5B08 What is the peak-to-peak voltage of a sine wave with an RMS voltage of 120.0 volts?
A. 84.8 volts
B. 169.7 volts
C. 240.0 volts
D. 339.4 volts

For sinusoidal signals, $V_{RMS} = V_{peak}/\sqrt{2}$. From this, the peak voltage is $V_{peak} = \sqrt{2}V_{RMS} = 1.41 \times 120\,V = 169.7\,V$. This is Answer B, but it is incorrect for this question. To find the peak-to-peak voltage, we multiply this result by 2. **Answer D** has the correct peak-to-peak voltage of 339.4 V.

G5B09 What is the RMS voltage of a sine wave with a value of 17 volts peak?
   A. 8.5 volts
   B. 12 volts
   C. 24 volts
   D. 34 volts

We use the formulas again for this question. For sinusoidal signals, $V_{RMS} = V_{peak}/\sqrt{2} = 17V/\sqrt{2} = 12V$. This makes **Answer B** the correct computation. Be careful with Answer A because it uses 2, and not $\sqrt{2}$. Answer C is $V_{peak} \times 2$, while Answer D is another math error.

G5B10 What percentage of power loss would result from a transmission line loss of 1 dB?
   A. 10.9 percent
   B. 12.2 percent
   C. 20.6 percent
   D. 25.9 percent

A loss of 1 dB corresponds to a factor of $10^{-1 \times 0.1} = 0.794$. Since dB is a power ratio, the signal has 79.4 % of its original power. This means that it has lost 20.6 % of the original power. **Answer C** is the correct choice.

G5B11 What is the ratio of peak envelope power to average power for an unmodulated carrier?
   A. 0.707
   B. 1.00
   C. 1.414
   D. 2.00

With an unmodulated carrier, the waveform is a pure sinusoid. In this case, the peak envelope power is the same as the RMS power, so the ratio is 1.00, as in **Answer B**. The other choices reflect misunderstandings of the definitions.

G5B12 What would be the RMS voltage across a 50 ohm dummy load dissipating 1200 watts?
A. 173 volts
B. 245 volts
C. 346 volts
D. 692 volts

Here, we need to invert the equation used earlier. The peak voltage, PEV, is computed using $PEV = \sqrt{2 \times R_L \times PEP} = \sqrt{2 \times 50 \times 1200} = 346V$. The peak envelope voltage is related to the RMS voltage by $V_{RMS} = 0.707 \, PEV = 245V$. This makes **Answer B** the right choice. Answer C is incorrect because it forgets the factor of $\sqrt{2}$ in converting from PEV to RMS voltage.

G5B13 What is the output PEP of an unmodulated carrier if an average reading wattmeter connected to the transmitter output indicates 1060 watts?
A. 530 watts
B. 1060 watts
C. 1500 watts
D. 2120 watts

This case is interesting because we do not need to make any calculations or conversions. The average power of an unmodulated carrier, as measured by the averaging wattmeter, is the PEP power. Therefore, the right choice is **Answer B**. The other answers are to see if you understand this basic principle.

G5B14 What is the output PEP from a transmitter if an oscilloscope measures 500 volts peak-to-peak across a 50 ohm resistive load connected to the transmitter output?
A. 8.75 watts
B. 625 watts
C. 2500 watts
D. 5000 watts

Using the same analysis, the peak envelope voltage is 1/2 the peak-to-peak voltage or 250 V. We square this quantity, and then divide the result by 2 times the 50-Ω load resistance or the PEP is $(250 \times 250) \div (2 \times 50)$ or 625 W. The correct choice is **Answer B**. The other choices have math errors.

5.5 G5C - Series and Parallel Circuits

5.5.1 Overview

The *Series and Parallel Circuits* question group in Subelement G5 examines the electrical principles for series and parallel circuits. The *Series and Parallel Circuits*

group covers topics such as
- Resistors, capacitors, and inductors in series and parallel
- Transformers

The test producer will select one of the 18 questions in this group for your exam.

5.5.2 Questions

G5C01 What causes a voltage to appear across the secondary winding of a transformer when an AC voltage source is connected across its primary winding?
A. Capacitive coupling
B. Displacement current coupling
C. Mutual inductance
D. Mutual capacitance

Because transformers deal with inductance, you should be able to eliminate Answers A and D since they involve capacitance. Mutual inductance, as given in **Answer C**, is the correct term for this effect. Answer B is to distract you.

G5C02 What happens if a signal is applied to the secondary winding of a 4:1 voltage step-down transformer instead of the primary winding?
A. The output voltage is multiplied by 4
B. The output voltage is divided by 4
C. Additional resistance must be added in series with the primary to prevent overload
D. Additional resistance must be added in parallel with the secondary to prevent overload

Answers B, C, and D are incorrect electronics, so they are not practical solutions. Reversing the order of applying the signal to the windings makes it a step-up instead of a step-down transformer, and **Answer A** is the right choice.

G5C03 Which of the following components increases the total resistance of a resistor?
A. A parallel resistor
B. A series resistor
C. A series capacitor
D. A parallel capacitor

Resisters add in series, so if we wish to make a larger resister, we add more resistance in series, as in **Answer B**. The other choices are incorrect electronics.

G5C04 What is the total resistance of three 100 ohm resistors in parallel?
  A. 0.30 ohms
  B. 0.33 ohms
  C. 33.3 ohms
  D. 300 ohms

When same-size resistors are in parallel, the total resistance is R/N where R is the individual resistor value, and N is the number of resistors combined. **Answer C** has the right value of $33.3\,\Omega$. The other choices represent math errors.

G5C05 If three equal value resistors in series produce 450 ohms, what is the value of each resistor?
  A. 1500 ohms
  B. 90 ohms
  C. 150 ohms
  D. 175 ohms

When same-size resistors are in series, the total resistance is $R \times N$ where R is the individual resistor value, and N is the number of resistors combined. **Answer C** has the right value of $150\,\Omega$. The other choices represent math errors.

G5C06 What is the RMS voltage across a 500-turn secondary winding in a transformer if the 2250-turn primary is connected to 120 VAC?
  A. 2370 volts
  B. 540 volts
  C. 26.7 volts
  D. 5.9 volts

This example is for a step-down voltage transformation (fewer turns on the secondary than on the primary). When we step the voltage down, the output voltage needs to be smaller than the input voltage, making Answers A and B incorrect. We compute the step-down ratio in terms of the ratio of the number of windings: $N_1/N_2 = 2250/500$ or 4.5. The voltage reduction will be $120V \div 4.5$ or 26.7 V as in **Answer C**. Answer D is incorrectly using 4.5^2 instead of 4.5.

G5C07 What is the turns ratio of a transformer used to match an audio amplifier having 600 ohm output impedance to a speaker having 4 ohm impedance?
  A. 12.2 to 1
  B. 24.4 to 1
  C. 150 to 1
  D. 300 to 1

When using the turns ratio to compute the resistance needed, we use the square root of the ratio. The ratio of $600\,\Omega$ to $4\,\Omega$ is 150, which is Answer C, which is incorrect. We still need to take the square root of this or 12.2, as in **Answer A**.

The other two answers have math mistakes.

G5C08 What is the equivalent capacitance of two 5.0 nanofarad capacitors and one 750 picofarad capacitor connected in parallel?
    A. 576.9 nanofarads
    B. 1733 picofarads
    C. 3583 picofarads
    D. 10.750 nanofarads

Capacitors in parallel add like resistors in series. In this case, $C_T = (5.0 + 5.0 + 0.750)$nF $= 10.750$ nF. This computation matches **Answer D**. The other choices are math mistakes.

G5C09 What is the capacitance of three 100 microfarad capacitors connected in series?
    A. 0.30 microfarads
    B. 0.33 microfarads
    C. 33.3 microfarads
    D. 300 microfarads

Capacitors in series add like resistors in parallel. In this case, $C_T = 1/(1/100 + 1/100 + 1/100)\mu F = 33.3\,\mu F$. This computation matches **Answer C**. Be careful with Answer D because it is for capacitors in parallel.

G5C10 What is the inductance of three 10 millihenry inductors connected in parallel?
    A. 0.30 henries
    B. 3.3 henries
    C. 3.3 millihenries
    D. 30 millihenries

Inductors in parallel add like resistors in parallel, so the total inductance is $L_T = 1/(1/L1 + 1/L2 + 1/L3) = 1/(1/0.01 + 1/0.01 + 1/0.01)$mH $= 3.3$ mH. This matches **Answer C**. Answer D is wrong because it is for inductors in series.

G5C11 What is the inductance of a 20 millihenry inductor connected in series with a 50 millihenry inductor?
    A. 0.07 millihenries
    B. 14.3 millihenries
    C. 70 millihenries
    D. 1000 millihenries

Inductors in series add like resistors in series, so the total inductance is given by $L_T = 20$ mH $+ 50$ mH $= 70$ mH. This corresponds to the computation in **Answer C**. Answer B is the result when the inductors are in parallel.

G5C12 What is the capacitance of a 20 microfarad capacitor connected in series with a 50 microfarad capacitor?
   A. 0.07 microfarads
   B. 14.3 microfarads
   C. 70 microfarads
   D. 1000 microfarads

The total capacitance for series capacitors is computed from $C_T = 1/(1/C1 + 1/C2) = 1/(1/20 + 1/50)\mu F = 14.3\,\mu F$. **Answer B** is the right computation. Answer C is the result for these capacitors connected in parallel.

G5C13 Which of the following components should be added to a capacitor to increase the capacitance?
   A. An inductor in series
   B. A resistor in series
   C. A capacitor in parallel
   D. A capacitor in series

We need to add capacitors to increase capacitance, so Answers A and B are incorrect. Capacitors in parallel add like resistors in series; therefore, to increase capacitance, we add more capacitors in parallel, as given in **Answer C**.

G5C14 Which of the following components should be added to an inductor to increase the inductance?
   A. A capacitor in series
   B. A resistor in parallel
   C. An inductor in parallel
   D. An inductor in series

To increase inductance, we need to add an inductor, and we can eliminate Answers A and B. To increase inductance, we add the inductors in series, making **Answer D** the correct method. Answer C will decrease the inductance.

G5C15 What is the total resistance of a 10 ohm, a 20 ohm, and a 50 ohm resistor connected in parallel?
   A. 5.9 ohms
   B. 0.17 ohms
   C. 10000 ohms
   D. 80 ohms

The total resistance will be computed from $R_T = 1/(1/R1 + 1/R2 + 1/R3) = 1/(1/10 + 1/20 + 1/50) = 5.9\Omega$. **Answer A** is the right choice. Answer D is for these resistors in series.

G5C16 Why is the conductor of the primary winding of many voltage step-up transformers larger in diameter than the conductor of the secondary winding?
- A. To improve the coupling between the primary and secondary
- B. To accommodate the higher current of the primary
- C. To prevent parasitic oscillations due to resistive losses in the primary
- D. To ensure that the volume of the primary winding is equal to the volume of the secondary winding

Because the primary is dealing with a larger voltage, you may correctly suspect that the primary also carries a larger current. A greater current demands a larger wire diameter to safely carry the current, as in **Answer B**.

G5C17 What is the value in nanofarads (nF) of a 22,000 picofarad (pF) capacitor?
- A. 0.22
- B. 2.2
- C. 22
- D. 220

Here we need to make a powers-of-ten conversion. There are $1000\,\text{pF}$ in $1\,\text{nF}$. In this case we divide by 1000, and $22\,000\,\text{pF} = 22\,\text{nF}$, as in **Answer C**. The other choices have wrong factors of 10.

G5C18 What is the value in microfarads of a 4700 nanofarad (nF) capacitor?
- A. 47
- B. 0.47
- C. 47,000
- D. 4.7

For this powers-of-ten conversion there are $1000\,\text{nF}$ in $1\,\mu\text{F}$. In this case we divide by 1000, and $4700\,\text{nF} = 4.76\,\mu\text{F}$, as in **Answer D**. The other choices have wrong factors of 10.

Chapter 6

G6 — CIRCUIT COMPONENTS

6.1 Introduction

Radio engineers use a variety of discrete components in their radio circuitry. These components may be discrete elements like resistors and capacitors, or semiconductor devices like transistors. This subelement covers the operating characteristics of these components, so you will have a better understanding of why they are in your circuits. As a refresher, Figure 6.1 contains common symbols for circuit components.

The *Circuit Components* subelement has the following question groups:

A. Discrete Elements
B. Integrated Circuits

Subelement 6 will generate two questions on the General Class examination.

6.2 Radio Engineering Concepts

Resistors and Capacitors The resistor questions cover two types:

Carbon Composite — general resistors that have moderate accuracy in their values, limited power dissipation potential, and are relatively inexpensive

Wire Wound — resistors composed of wound wire that can withstand higher temperatures and currents than carbon resistors; in certain respects, they also behave like inductors, especially at high frequencies

Most resistive devices are also temperature sensitive; their resistance value changes with temperature. When this is the desired behavior, we have a thermometer. We need to include temperature compensation techniques to stabilize performance when the temperature sensitivity is undesired. The *temperature coefficient* of the resistor determines the degree of change with temperature.

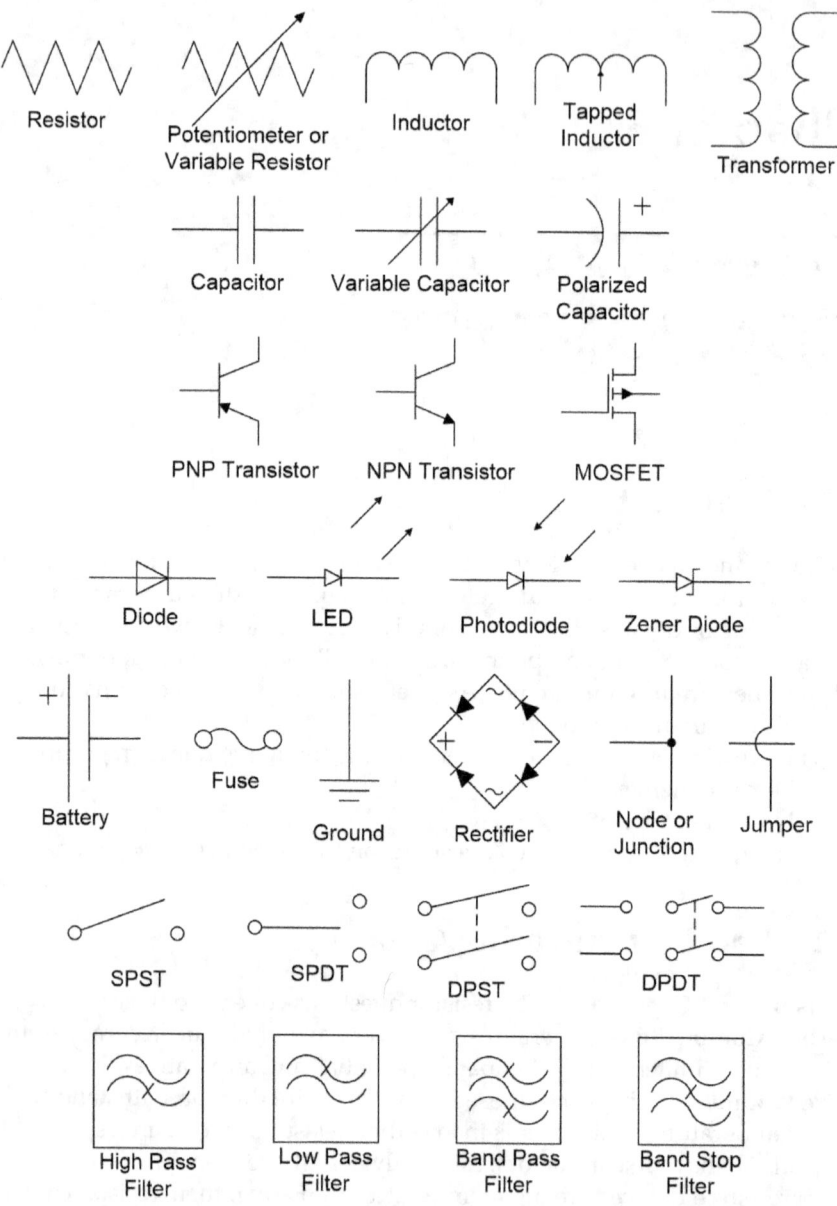

Figure 6.1: Common symbols used for circuit elements.

Table 6.1: Battery depth of discharge per cell.

Type	Fully Charged (V/cell)	Discharge Depth (V/cell)
NiCd	1.2 V	1.0 V
Li-ion	3.6 V	3.0 V
Lead Acid	2.0 V	1.75 V

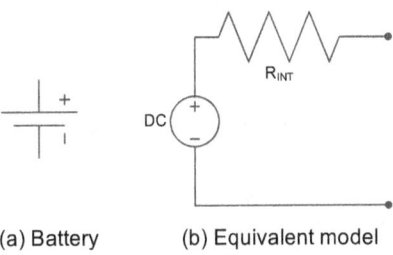

(a) Battery　　　(b) Equivalent model

Figure 6.2: Circuit equivalent model for a battery.

The capacitor questions cover two technology types as well:

Ceramic — capacitors using a solid dielectric material that is inexpensive to manufacture

Electrolytic — capacitors with a liquid material internally having a high capacitance

Capacitors can be either electrically *polarized* with a designated terminal polarity or *non-polarized*. Ceramic capacitors are generally non-polarized, while electrolytic capacitors are often polarized.

Batteries　Batteries typically have a maximum rated charge per cell. The user discharges the battery with use to power devices. However, for rechargeable batteries, the user should not completely drain the battery if there is a desire to have many recharge cycles. Table 6.1 shows rule-of-thumb estimates for fully-charged and discharged cell voltages for several common rechargeable battery types. Notice that you should not completely discharge the battery.

Figure 6.2 illustrates an equivalent circuit model for a battery. The battery model is composed of a fixed Direct Current (DC) source in series with an internal resistance, R_{INT}. This resistor reduces the output voltage, and restricts the current available to the load. Naturally, this should be as small as possible.

Diodes and Transistors　Diodes are devices to allow current flow in a single direction. When they are *forward biased*, they permit current flow, and when *reverse biased*, they prevent current flow. Table 6.2 lists several common types

Table 6.2: Common diode types.

Type	Junction Voltage	Notes
Si	0.7 V	based on silicon
Ge	0.3 V	based on germanium
Schottky	0.2 V	rapid switching
LED	1.4 to 4 V	emits light when forward biased

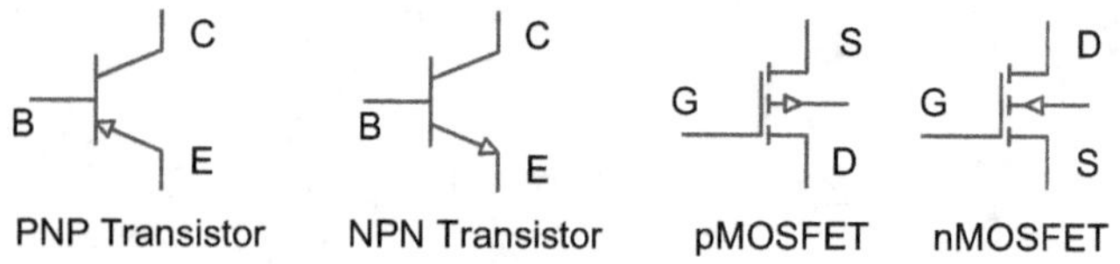

Figure 6.3: BJT and MOSFET transistor types and the associated schematic symbols with the connector labels.

of semiconductor diodes that you need to know about for the General Class examination.

Two general transistor technologies appear in the license examination questions: the Bipolar Junction Transistor (BJT) and the Metal Oxide Semiconductor Field Effect Transistor (MOSFET). The type of semiconductor material the designer uses in fabricating the transistor controls the transistor's exact characteristics. There are two material types used: *n-type* which has excess electrons for conducting electricity, and *p-type* which has an excess of positive material. Figure 6.3 illustrates the schematic symbols for the BJT and MOSFET transistors. The connectors for the BJT are Emitter (E), Collector (C), and Base (B), while the connectors for the MOSFET are Source (S), Drain (D), and Gate (G).

Vacuum Tubes   Vacuum tubes have been around for over a century now, with the triode being the earliest design. While transistors have replaced most vacuum tubes in the small-signal market, vacuum tubes still hold their own in the high-amplification market. The major electrical elements of a triode vacuum tube are

Cathode — the electron emitter

Anode — the electron collector

Gate — controls the flow of the electrons between the cathode and the anode

Heater — heats the cathode to make the electron emitter more efficient

The triode works by having the heater warm the cathode and lowering the barrier to electron emission. The voltage potential between the cathode and the anode pulls electrons towards the anode. Designers use the voltage across the gate to control the electron flow.

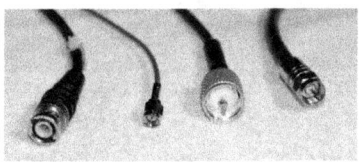

(a) BNC, SMA, UHF (PL-259), and F connectors (left to right).

(b) Various microphone DIN connectors.

(c) RCA, ¼-inch, and 3.5 mm phone plugs (left to right).

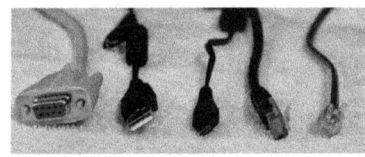

(d) 9-pin serial (DE-9), USB, mini USB, networking (RJ45), and telephone (RJ11) plugs (left to right).

Figure 6.4: Pictures of connectors used between rigs and support devices in the amateur radio shack.

Connectors   In the Technician Class study guide, we saw a variety of electronic connectors for use in the shack. As a review, we group the cable and connector functions into four general categories

RF Connectors — connectors used to carry Radio Frequency (RF) signals between the rig, the antenna, and any intermediate components

Microphone Connectors — connectors to attach a microphone to the rig

Audio Connectors — connectors to attach speakers and headphones to the rig

Data Connectors — connectors to allow the rig to interface with a computer

Figure 6.4 illustrates many of the common connector types that can be found in an amateur radio shack. This is an evolving mix of connectors. With more devices supporting Wi-Fi and Bluetooth wireless connections, many of these non-RF connectors may disappear in the future.

6.3   G6A - Discrete Elements

6.3.1   Overview

The *Discrete Elements* question group in Subelement G6 quizzes you on the electrical properties of discrete analog devices. The *Discrete Elements* group covers topics such as

- Resistors, capacitors, and inductors
- Rectifiers and solid state diodes and transistors
- Vacuum tubes
- Batteries

The test producer will select one of the 14 questions in this group for your exam.

6.3.2 Questions

G6A01 What is the minimum allowable discharge voltage for maximum life of a standard 12 volt lead-acid battery?

A. 6 volts
B. 8.5 volts
C. 10.5 volts
D. 12 volts

Batteries designed to be recharged often, like lead-acid batteries, have a maximum depth of discharge rating to permit them to have many charge/discharge cycles. This question is asking about general operating conditions for standard 12-V batteries. The battery has six cells so, from Table 6.2, the maximum discharge should be $6 \times 1.75\,\text{V} = 10.5\,\text{V}$, as in **Answer C**. Going to 6 V or 8.5 V is possible, but those levels will not maintain the maximum life of the battery. Answer D represents never discharging, which makes the battery useless.

G6A02 What is an advantage of the low internal resistance of nickel-cadmium batteries?

A. Long life
B. High discharge current
C. High voltage
D. Rapid recharge

The internal resistance of a battery, as Figure 6.2 shows, will limit the ability to supply high current, so **Answer B** is the right choice. It will not extend life, increase the voltage, or allow a rapid recharge, so do not choose those answers.

G6A03 What is the approximate junction threshold voltage of a germanium diode?

A. 0.1 volt
B. 0.3 volts
C. 0.7 volts
D. 1.0 volts

Junction threshold voltages may be items you need to memorize. For germanium diodes, the voltage is 0.3 V, as in **Answer B**. Be careful because we find 0.7 V in many silicon diodes. The other choices are there to distract you.

G6A04 Which of the following is an advantage of an electrolytic capacitor?
A. Tight tolerance
B. Much less leakage than any other type
C. High capacitance for a given volume
D. Inexpensive RF capacitor

Electrolytic capacitors are known for having a relatively high capacitance for a given volume, as in **Answer C**, which is why designers use them in many electronic circuits. The other choices are electrically incorrect.

G6A05 What is the approximate junction threshold voltage of a conventional silicon diode?
A. 0.1 volt
B. 0.3 volts
C. 0.7 volts
D. 1.0 volts

This question brings us back to diode voltages. We saw that Answer B was for germanium diodes, so it is not the right choice here. The silicon diodes need 0.7 V, making **Answer C** the right choice. Answers A and D are to distract you.

G6A06 Which of the following is a reason not to use wire-wound resistors in an RF circuit?
A. The resistor's tolerance value would not be adequate for such a circuit
B. The resistor's inductance could make circuit performance unpredictable
C. The resistor could overheat
D. The resistor's internal capacitance would detune the circuit

A wire-wound resistor can have significant inductive properties (remember the transformer from Chapter 5). A RF circuit is often a tuned circuit, defined by its capacitors and inductors. The wire-wound resistor will have different impedance values with frequency, so the performance will be more challenging to predict, which makes **Answer B** the right choice. Answer D is incorrect because the device has inductance properties. We are not concerned with Answers A and C with properly-chosen components.

G6A07 What are the stable operating points for a bipolar transistor used as a switch in a logic circuit?
A. Its saturation and cutoff regions
B. Its active region (between the cutoff and saturation regions)
C. Its peak and valley current points
D. Its enhancement and depletion modes

Bipolar transistor switches are run to either saturation or cutoff to represent the

switch states, so **Answer A** is the right choice. Answer B is the desired amplifier region, not the switch-operating mode. Answers C and D are to distract you.

G6A08 What is an advantage of using a ferrite core toroidal inductor?
   A. Large values of inductance may be obtained
   B. The magnetic properties of the core may be optimized for a specific range
      of frequencies
   C. Most of the magnetic field is contained in the core
   D. All of these choices are correct

Since each of the effects mentioned in Answers A, B, and C is true, **Answer D** is
the best choice.

G6A09 Which of the following describes the construction of a MOSFET?
   A. The gate is formed by a back-biased junction
   B. The gate is separated from the channel with a thin insulating layer
   C. The source is separated from the drain by a thin insulating layer
   D. The source is formed by depositing metal on silicon

The question does not specify the exact type
of MOSFET, so we will use Figure 6.5 as
a typical representation. As you can see, the
gate is separated from the p-channel by a
layer of silicon dioxide, which is an insulator. This geometry makes **Answer B** the
correct choice. From the figure, we can see
that Answer C does not match the geometry. Answer A is for a Junction Field Effect Transistor (JFET), and Answer D does
not correspond to the MOSFET construction
method.

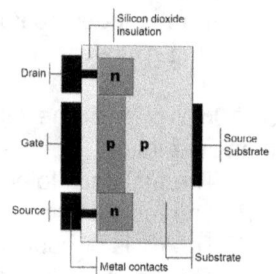

Figure 6.5: The MOS-
FET construction configu-
ration.

G6A10 Which element of a triode vacuum tube is used to regulate the flow of
electrons between cathode and plate?
   A. Control grid
   B. Heater
   C. Screen Grid
   D. Trigger electrode

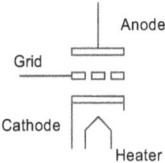

The triode vacuum tube is illustrated in Figure 6.6. The control grid is the regulator of electron flow, so **Answer A** is the right choice. While the heater is part of the tube, it helps to liberate electrons from the cathode, so that is incorrect here. The other elements mentioned do not control electron flow or are not part of the triode.

Figure 6.6: The circuit symbol for a triode.

G6A11 What happens when an inductor is operated above its self-resonant frequency?
  A. Its reactance increases
  B. Harmonics are generated
  C. It becomes capacitive
  D. Catastrophic failure is likely

If the inductor operates above its self-resonant frequency, then it becomes capacitive, as in **Answer C**. The other choices are electrically incorrect.

G6A12 What is the primary purpose of a screen grid in a vacuum tube?
  A. To reduce grid-to-plate capacitance
  B. To increase efficiency
  C. To increase the control grid resistance
  D. To decrease plate resistance

Manufacturers added the screen grid to vacuum tubes to reduce the grid-to-plate capacitance, so **Answer A** is the right choice. The other options are to distract you.

G6A13 Why is the polarity of applied voltages important for polarized capacitors?
  A. Incorrect polarity can cause the capacitor to short-circuit
  B. Reverse voltages can destroy the dielectric layer of an electrolytic capacitor
  C. The capacitor could overheat and explode
  D. All of these choices are correct

If we apply the wrong polarity to a polarized capacitor, we may discover that one possible result is an effect listed in Answers A, B, or C. That makes **Answer D** the best choice for this question.

G6A14 Which of the following is an advantage of ceramic capacitors as compared to other types of capacitors?
A. Tight tolerance
B. High stability
C. High capacitance for given volume
D. Comparatively low cost

Ceramic capacitors generally have low tolerance, low stability, and low capacitance per volume. However, they are inexpensive, so **Answer D** is the correct choice. The high capacitance in Answer C is for electrolytic capacitors, so be careful with that choice.

6.4   G6B - Integrated Circuits

6.4.1   Overview

The *Integrated Circuits* question group in Subelement G6 tests you on digital devices, Integrated Circuits (ICs), and connectors. The *Integrated Circuits* group covers topics such as
- Analog and digital integrated circuits (ICs)
- Microwave ICs (MMICs)
- Microprocessors and memory
- I/O devices, display devices, connectors, and ferrite cores

The test producer will select one of the 13 questions in this group for your exam.

6.4.2   Questions

G6B01 What determines the performance of a ferrite core at different frequencies?
A. Its conductivity
B. Its thickness
C. The composition, or "mix," of materials used
D. The ratio of outer diameter to inner diameter

A ferrite core is made by carefully selecting materials for its mix to achieve specific performance goals. This result indicates that **Answer C** is the right choice. The other answers do not affect the frequency performance of the core. Ferrite "mixture 31" is commonly used in amateur applications.

G6B02 What is meant by the term MMIC?
A. Multi-Megabyte Integrated Circuit
B. Monolithic Microwave Integrated Circuit
C. Military Manufactured Integrated Circuit
D. Mode Modulated Integrated Circuit

This acronym is for a Monolithic Microwave Integrated Circuit (MMIC), which makes **Answer B** the right choice. The other options are distractions to confuse you.

G6B03 Which of the following is an advantage of CMOS integrated circuits compared to TTL integrated circuits?
A. Low power consumption
B. High power handling capability
C. Better suited for RF amplification
D. Better suited for power supply regulation

A significant reason Complementary Metal Oxide Semiconductor (CMOS) has overtaken Transistor-Transistor Logic (TTL) is because of its relatively low power consumption. **Answer A** is the correct choice. The other options are not necessarily valid for CMOS.

G6B04 What is meant by the term ROM?
A. Resistor Operated Memory
B. Read Only Memory
C. Random Operational Memory
D. Resistant to Overload Memory

You should be able to spot Read Only Memory (ROM) in **Answer B** as the correct answer. The others are just distractions.

G6B05 What is meant when memory is characterized as non-volatile?
A. It is resistant to radiation damage
B. It is resistant to high temperatures
C. The stored information is maintained even if power is removed
D. The stored information cannot be changed once written

Non-volatile memory retains its contents when the user removes the power, so **Answer C** is the correct choice. The other choices do not describe non-volatile memory.

G6B06 What kind of device is an integrated circuit operational amplifier?
A. Digital
B. MMIC
C. Programmable Logic
D. Analog

The operational amplifier is an analog IC, so **Answer D** is correct. It is not digital, so Answer A is eliminated. It is not programmable, so we also eliminate Answer C. A MMIC is a multi-component IC module, not just a single amplifier.

G6B07 Which of the following describes a type N connector?
A. A moisture-resistant RF connector useful to 10 GHz
B. A small bayonet connector used for data circuits
C. A threaded connector used for hydraulic systems
D. An audio connector used in surround-sound installations

A Type-N connector is described in **Answer A**, so this is the right choice. This connector is for RF, not audio, so we eliminate so Answer D. We do not use hydraulics in RF connectors, so Answer C is incorrect. Answer B is close to the description of a BNC connector, so it is not a good choice here.

G6B08 How is an LED biased when emitting light?
A. Beyond cutoff
B. At the Zener voltage
C. Reverse Biased
D. Forward Biased

A Light Emitting Diode (LED) "turns on" and emits light when it is forward biased, which makes **Answer D** the correct choice. LEDs work like regular diodes when in reverse bias mode and do not emit light. Zener diodes are a particular category, and designers do not use them as LEDs.

G6B09 Which of the following is a characteristic of a liquid crystal display?
A. It utilizes ambient or back lighting
B. It offers a wide dynamic range
C. It consumes relatively high power
D. It has relatively short lifetime

A Liquid Crystal Display (LCD) does not require high power, does not have a wide dynamic range, nor does it have a short lifetime, so Answers B, C, and D are incorrect. The LCD typically uses backlighting, so **Answer A** is the correct choice.

G6B10 How does a ferrite bead or core reduce common-mode RF current on the shield of a coaxial cable?
A. By creating an impedance in the current's path
B. It converts common-mode current to differential mode
C. By creating an out-of-phase current to cancel the common-mode current
D. Ferrites expel magnetic fields

The ferrite material makes an impedance to the RF current, as in **Answer A**. The other choices are electrically incorrect statements.

G6B11 What is a type SMA connector?
   A. A large bayonet connector usable at power levels in excess of 1 KW
   B. A small threaded connector suitable for signals up to several GHz
   C. A connector designed for serial multiple access signals
   D. A type of push-on connector intended for high voltage applications

SMA connectors are small, threaded connectors, as in Figure 6.4a, used in RF applications up to several GHz, as in **Answer B**. The other descriptions are not for SMA.

G6B12 Which of these connector types is commonly used for audio signals in Amateur Radio stations?
   A. PL-259
   B. BNC
   C. RCA Phono
   D. Type N

The RCA phono connector shown in Figure 6.4c is the only audio connector in the list, so **Answer C** is the correct choice. The other three connectors are for RF, so they are not commonly used for audio connections.

G6B13 Which of these connector types is commonly used for RF connections at frequencies up to 150 MHz?
   A. Octal
   B. RJ-11
   C. PL-259
   D. DB-25

Figure 6.4 shows many of these connectors. The DB-25 in Answer D is a computer serial data cable, so it is not good for RF. The RJ-11 is a telephone connector, so it is also not a good choice for an RF connection. Designers use the octal connector in Answer A to hold vacuum tubes in circuits, which is also incorrect. The PL-259 connector of **Answer C** is the right choice. Note: the PL-259 connector is also called a UHF connector. Even though it is called a UHF connector, users do not often use it on the Ultra High Frequency (UHF) band. This name is an old (WWII vintage) designation.

Chapter 7

G7 — PRACTICAL CIRCUITS

7.1 Introduction

The *Practical Circuits* subelement builds on the previous study areas of both the Technician Class and General Class study guides to look at complex components such as logic gates, power supplies, and block diagrams for radios. We will use the circuit symbols in Figure 6.1 here again. The questions will look at the operating characteristics of these selected circuits and electronic elements. The *Practical Circuits* subelement has the following question groups:

    A. Power Supplies
    B. Digital, Amplifier, and Oscillator Circuits
    C. RF Circuits

Subelement 7 will generate three questions on the General Class examination.

7.2 Radio Engineering Concepts

Filters   Earlier, in Chapter 4, we saw the fundamental filter types in Figure 4.3. In this chapter, we will need to go into more detail. Designers characterize filters by key points and measures such as

Cutoff Frequency — the frequency where the attenuation reaches 3 dB from the 0-dB reference

Roll Off — the rate at which the frequency attenuation increases after the cutoff frequency

Pass Band — the range of frequencies where the attenuation is less than 3 dB

Stop Band — the range of frequencies where the attenuation exceeds a specified value such as 26 dB

Figure 7.1 shows these in a Low Pass Filter (LPF) example. Similar measures exist for the other filter types.

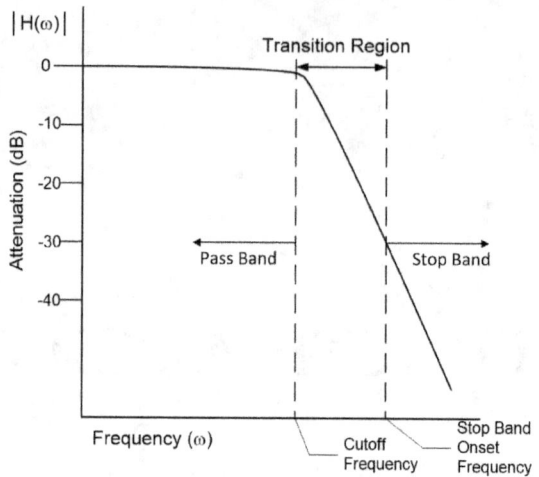

Figure 7.1: Key regions for a low-pass filter.

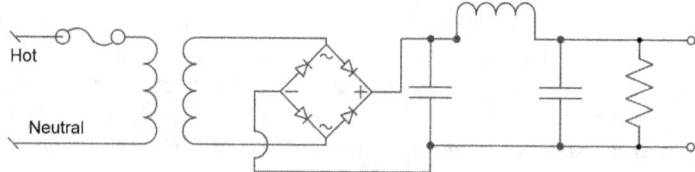

Figure 7.2: Representative circuit for a DC power supply.

Power Supplies   Direct Current (DC) power supplies are fundamental parts of radio equipment. Figure 7.2 illustrates a typical circuit for converting Alternating Current (AC) input power to a constant-voltage DC supply. The circuit uses a transformer to change the 120-V AC input to a useful DC level, such as 12 V. The output of the transformer is connected to a *bridge rectifier*. This is a diode circuit that makes the AC voltage into a positive-only voltage. After the rectifier is an Inductor-Capacator (*LC*) circuit that smooths the AC ripples and makes the constant DC voltage. After the filter is a resistor to bleed off the stored charge in the capacitors when the AC source is removed. This is to prevent a shock hazard from the stored charge.

Half and Full-Wave Rectifiers   In Figure 7.2 we saw a diode rectifier circuit to remove the negative polarity parts of the AC input. There are two different types of diode rectifier circuits: *half-wave* and *full-wave* rectification. Figure 7.3 shows both configurations. With the half-wave rectifier, the circuit eliminates the negative-polarity pulses from the input AC source and only passes the positive-polarity pulses to the filter network. The diode goes into reverse bias during those periods and blocks the current flow, which gives an output over

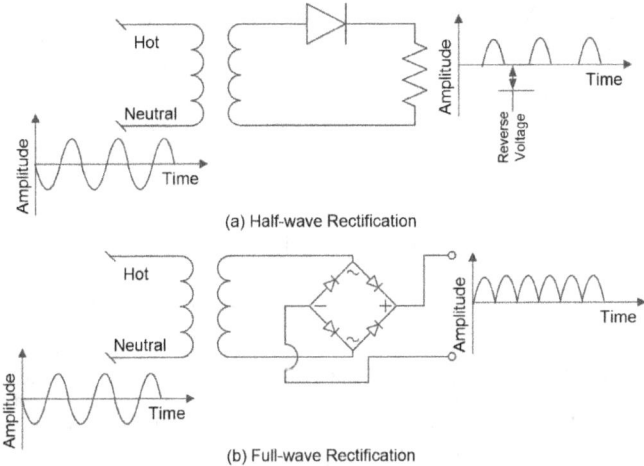

Figure 7.3: Half-wave and full-wave rectification.

Table 7.1: Combinational logic functions

Input 1	Input 2	AND	NAND	OR	NOR
0	0	0	1	0	1
0	1	0	1	1	0
1	0	0	1	1	0
1	1	1	0	1	0

only 180° of the full 360° of the AC input.

The full-wave rectifier is more efficient at the cost of a few extra diodes. Here, the negative-polarity pulses are flipped into positive-polarity pulses by the diode arrangement, and we have output pulses over the full 360° of the AC input.

Logic Functions   You need to be familiar with several fundamental combinatorial logic functions for the General Class exam. They describe the possible output states for combinations of possible input states. Each function uses two inputs, either a 0 or a 1. Table 7.1 lists the outputs for the logic functions with these inputs. The table is for two inputs, but one can extend it for more inputs. Figure 7.4 illustrates common logic gate circuit symbols.

RF Amplifiers   There are many types of Radio Frequency (RF) amplifiers available for boosting the output power of the transmitter's signal. The General Class examination quizzes you on the names and characteristics of these amplifiers. Table 7.2 lists the standard amplifier classes and the features you need to

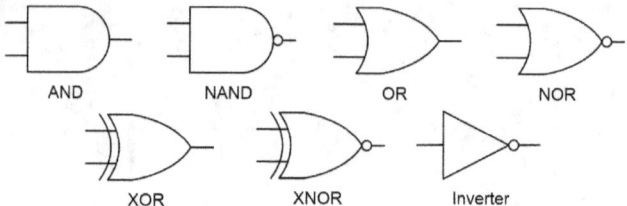

Figure 7.4: Common symbols used for logic gates.

Table 7.2: Characteristics of RF amplifiers

Class	Definition	Efficiency	Characteristics
A	Operate over full 360° input angle	25 % to 30 %	linear operation
AB	Two amplifiers; each operates over 180° input angle	50 % to 60 %	mostly linear operation
B	Two amplifiers operate over 180° input angle	up to 65 %	acceptable linearity
C	Operate near 90° input angle	up to 80 %	poor linearity

know. The Class A amplifier is the most linear but least power efficient. The Class AB and Class B amplifiers use two amplification segments to cover the entire 360° carrier cycle. The Class C amplifier is the most non-linear and the most efficient of the group.

SSB Transmitters   Single Sideband (SSB) transmissions can be made several ways using either analog components or Digital Signal Processor (DSP) methods. For the General Class examination, the question pool designers are assuming a SSB transmitter using the components shown in Figure 7.5. The transmitter's balanced modulator takes the phone input from the microphone and produces Dual Sideband (DSB) with it. The carrier frequency for the *Balanced Modulator* is chosen based on if Lower Side Band (LSB) or Upper Side Band (USB) is the desired mode. The *Filter* removes the undesired sideband to produce SSB. The *Mixer* then translates the SSB signal to the desired RF transmission frequency. The RF amplifier and antenna then emit the signal to the receiving station.

CW Receivers   Continuous Wave (CW) has a long tradition in amateur radio, and Figure 7.6 illustrates a typical CW receiver process. The RF signal enters through the antenna, is filtered to remove noise, and amplified to boost the signal's strength. Next, it is mixed with a local oscillator to shift the signal from

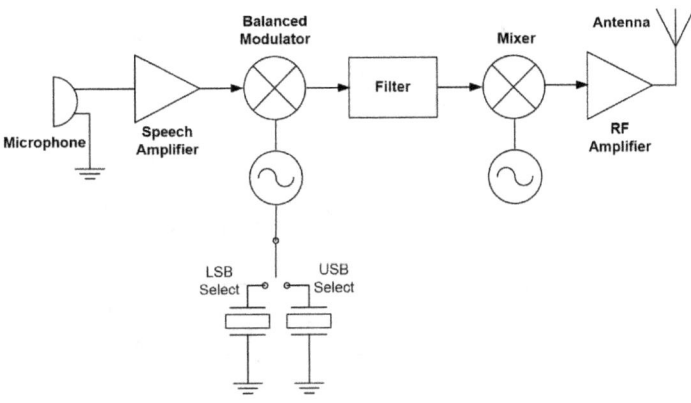

Figure 7.5: Example SSB transmitter configuration.

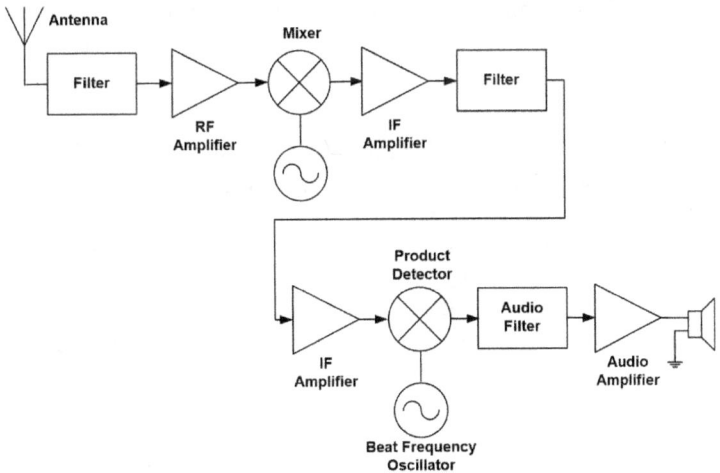

Figure 7.6: Example CW receiver configuration.

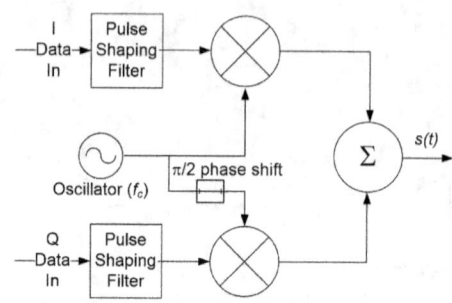

Figure 7.7: Digital PSK transmission.

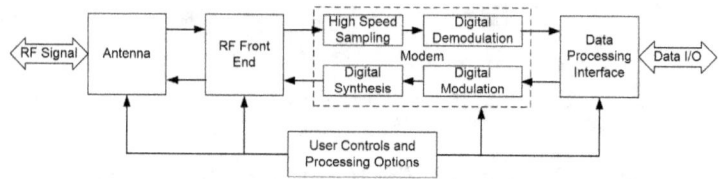

Figure 7.8: Software Defined Radio block diagram.

the RF carrier location to the receiver's standard Intermediate Frequency (IF) frequency. That IF signal is next filtered and amplified to prepare it for detection. Since this is CW, the IF signal will be a tone above the audible range that the circuit turns on and off to represent the dots and dashes of the CW code. The product detector multiplies the IF signal with the Beat Frequency Oscillator (BFO) to produce an audible tone. This tone is filtered, amplified, and sent to the speaker.

Digital Radios   Transmitting digital data, such as PSK31, builds on analog transmission techniques. Figure 7.7 shows how we generate a digital Phase Shift Keying (PSK) transmission by using Amplitude Modulation (AM) techniques to modulate the carrier with the digital data. In this figure, we separate the bits into "I" and "Q" channels to transmit up to two bits at a time. Notice that the oscillator uses a 90° or $\pi/2$ phase shift between the channels to keep the individual carrier signals from interfering. This transmitter generates Quadrature Phase Shift Keying (QPSK) when it uses both input data channels or Binary Phase Shift Keying (BPSK) when it uses only one channel.

Designers originally built the transmitter in Figure 7.7 with individual analog components. In modern radios, designers have replaced much of the analog signal processing with DSP components that make the inside of the radio look more like the inside of a computer system. Figure 7.8 illustrates the Software Defined Radio (SDR) block diagram with the DSP components inside the dashed box. On the transmission side, the DSP controls the modulation format and any necessary filtering. On the reception side, the Direct Digital Synthesis (DDS) components permit sampling of the RF waveform and immediate signal processing. The DSP software handles all of the demodulation processing. The user has great flexibility in controlling the SDR program options.

7.3   G7A - Power Supplies

7.3.1   Overview

The *Power Supplies* question group in Subelement G7 tests you over the principles of power supply operations and recognizing circuit symbols. The test producer will select one of the 13 questions in this group for your exam.

7.3.2   Questions

G7A01 What useful feature does a power supply bleeder resistor provide?
A. It acts as a fuse for excess voltage
B. It ensures that the filter capacitors are discharged when power is removed
C. It removes shock hazards from the induction coils
D. It eliminates ground loop current

Answers B and C seem to be the best candidates if you read the question carefully since they deal with safety issues. If your resistor acts like a fuse, (1) you will have let the magic white smoke out, and it will no longer work, and (2) using devices other than fuses as a fuse is a bad idea. The electrical purpose of the bleeder resistor is to discharge capacitors, as given in **Answer B**, so that is the correct choice. Figure 7.2 shows the bleeder resistor and the capacitor. Answers C and D are not true because they are electrically incorrect statements.

G7A02 Which of the following components are used in a power supply filter network?
A. Diodes
B. Transformers and transducers
C. Quartz crystals
D. Capacitors and inductors

As we saw in Figure 7.2, power supply filters are built with capacitors and inductors, so **Answer D** is the correct choice. Diodes do not provide filtering action, so Answer A is incorrect. Transistors are also not used in filters, so Answer B is out. Crystals can be used in signal processing filters but not in power supply filters, so Answer C is not a correct choice either.

G7A03 Which type of rectifier circuit uses two diodes and a center-tapped transformer?
  A. Full-wave
  B. Full-wave bridge
  C. Half-wave
  D. Synchronous

Figure 7.3(a) shows that a half-wave rectifier uses one diode, making Answer C incorrect. Full-wave rectification requires either two or four diodes, depending upon the circuit. Figure 7.3(b) shows a full-wave rectifier with a diode bridge circuit containing four diodes. This figure corresponds to Answer B, which is incorrect because there is no transformer. If you use a center-tapped transformer, you can replace the diode bridge with two diodes, and you correctly have the full-wave rectifier of **Answer A**. Answer D is to distract you.

G7A04 What is an advantage of a half-wave rectifier in a power supply?
  A. Only one diode is required
  B. The ripple frequency is twice that of a full-wave rectifier
  C. More current can be drawn from the half-wave rectifier
  D. The output voltage is two times the peak output voltage of the transformer

Figure 7.3 shows that the half-wave rectifier requires only one diode while a full-wave rectifier requires more diodes. Diode count is the advantage for this question, and **Answer A** is the right choice. The other answers are electrically incorrect.

G7A05 What portion of the AC cycle is converted to DC by a half-wave rectifier?
  A. 90 degrees
  B. 180 degrees
  C. 270 degrees
  D. 360 degrees

A half-wave rectifier circuit conducts over ½ of a cycle or 180°. This behavior makes **Answer B** the right choice. Answer D is for the full-wave rectifier, so be careful.

G7A06 What portion of the AC cycle is converted to DC by a full-wave rectifier?
  A. 90 degrees
  B. 180 degrees
  C. 270 degrees
  D. 360 degrees

A full-wave rectifier circuit conducts over the full cycle or 360°, so **Answer D** is the correct choice. Answer B is for the half-wave rectifier.

G7A07 What is the output waveform of an unfiltered full-wave rectifier connected to a resistive load?
- A. A series of DC pulses at twice the frequency of the AC input
- B. A series of DC pulses at the same frequency as the AC input
- C. A sine wave at half the frequency of the AC input
- D. A steady DC voltage

From Figure 7.3(b) for the full-wave rectifier circuit, we can see that the output has a series of pulses at twice the input frequency. This waveform makes **Answer A** the right choice.

G7A08 Which of the following is an advantage of a switchmode power supply as compared to a linear power supply?
- A. Faster switching time makes higher output voltage possible
- B. Fewer circuit components are required
- C. High frequency operation allows the use of smaller components
- D. All of these choices are correct

Choices A and B are electrically incorrect statements for switched-mode supplies, making Answer D wrong as well. The small-sized components of **Answer C** are correct for switched-mode power supplies.

G7A09 Which symbol in figure G7-1 represents a field effect transistor?
- A. Symbol 2
- B. Symbol 5
- C. Symbol 1
- D. Symbol 4

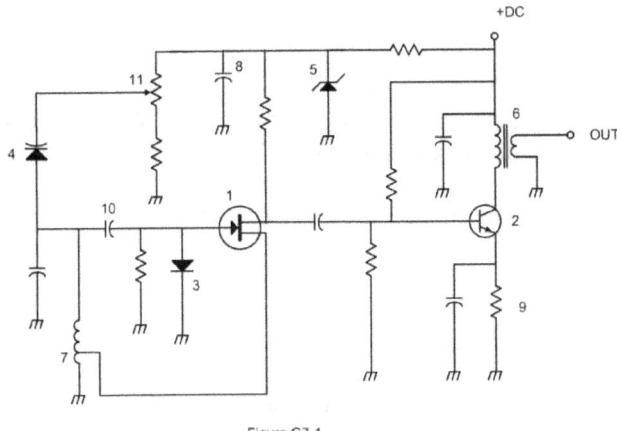

Figure G7-1

Figure 7.9: Figure G-1 for the General Class question pool.

Using Figure 6.1 as a guide, symbol 2 is an NPN transistor, symbol 5 is a Zener diode, symbol 1 is a Field Effect Transistor (FET), and symbol 4 is a varactor diode. **Answer C** is the correct choice.

G7A10 Which symbol in figure G7-1 represents a Zener diode?
A. Symbol 4
B. Symbol 1
C. Symbol 11
D. Symbol 5

Symbol 4 is the varactor diode, symbol 1 is the FET, symbol 11 is a variable resistor (also called a potentiometer), and symbol 5 is a Zener diode. Be sure to select **Answer D** as the correct choice.

G7A11 Which symbol in figure G7-1 represents an NPN junction transistor?
A. Symbol 1
B. Symbol 2
C. Symbol 7
D. Symbol 11

Are you catching on to the symbols? Symbol 1 is the FET, symbol 2 is the NPN transistor, symbol 7 is a tapped inductor, and symbol 11 is a variable resistor. Choose **Answer B** for the NPN transistor.

G7A12 Which symbol in Figure G7-1 represents a solid core transformer?
A. Symbol 4
B. Symbol 7
C. Symbol 6
D. Symbol 1

Symbol 4 is the varactor, symbol 7 is the tapped inductor, symbol 6 is the multiple-winding transformer, and symbol 1 is the FET. Did you recognize symbol 6 as the new symbol for the transformer and **Answer C** as the correct choice?

G7A13 Which symbol in Figure G7-1 represents a tapped inductor?
A. Symbol 7
B. Symbol 11
C. Symbol 6
D. Symbol 1

You should be able to get this one right away. Symbol 7 is the tapped inductor, so **Answer A** is the right choice. Did you remember that symbol 11 is the variable resistor, symbol 6 is the multiple-winding transformer, and symbol 1 is the FET?

7.4 G7B - Digital, Amplifier, and Oscillator Circuits

7.4.1 Overview

The *Digital, Amplifier, and Oscillator Circuits* question group in Subelement G7 tests you over a variety of circuits used in radio electronics. The test producer will select one of the 11 questions in this group for your exam.

7.4.2 Questions

G7B01 What is the reason for neutralizing the final amplifier stage of a transmitter?
A. To limit the modulation index
B. To eliminate self-oscillations
C. To cut off the final amplifier during standby periods
D. To keep the carrier on frequency

If properly designed, the neutralizing circuitry will not affect the modulation characteristics of the signal as in Answers A and D. Designers do not use neutralization for the purpose given in Answer C. Amplifiers can suffer from self-oscillations that produce spurious signals, and eliminating them is the function of the circuit, as in **Answer B**.

G7B02 Which of these classes of amplifiers has the highest efficiency?
A. Class A
B. Class B
C. Class AB
D. Class C

Generally, the amplifier efficiency is lowest in Class A and gets higher as we move towards Class C. **Answer D** is the correct choice.

G7B03 Which of the following describes the function of a two-input AND gate?
A. Output is high when either or both inputs are low
B. Output is high only when both inputs are high
C. Output is low when either or both inputs are high
D. Output is low only when both inputs are high

Table 7.1 shows the output for an AND gate given the possible input states. If you look at the choices given, only **Answer B** corresponds to the table. The output state is 1 only if both inputs are a 1.

G7B04 Which of the following describes the function of a two input NOR gate?
    A. Output is high when either or both inputs are low
    B. Output is high only when both inputs are high
    C. Output is low when either or both inputs are high
    D. Output is low only when both inputs are high

Table 7.1 shows the output for a NOR gate given the possible input states. If you look at the choices given, only **Answer C** corresponds to the table. The output state is 0 if either input is a 1.

G7B05 How many states does a 3-bit binary counter have?
    A. 3
    B. 6
    C. 8
    D. 16

The "3-bit" refers to the number of bits necessary to hold the count value. The number of states is $2^3 = 8$. This result makes **Answer C** the correct computation.

G7B06 What is a shift register?
    A. A clocked array of circuits that passes data in steps along the array
    B. An array of operational amplifiers used for tri-state arithmetic operations
    C. A digital mixer
    D. An analog mixer

A shift register is neither an analog nor a digital mixer, so we eliminate Answers C and D. A shift register does not use operational amplifiers, so Answer B is incorrect. The shift register is a clocked data array as in **Answer A**.

G7B07 Which of the following are basic components of a sine wave oscillator?
    A. An amplifier and a divider
    B. A frequency multiplier and a mixer
    C. A circulator and a filter operating in a feed-forward loop
    D. A filter and an amplifier operating in a feedback loop

There are many types of oscillators, but the one thing they all have in common is a form of feedback loop that makes the circuit oscillate. This design makes **Answer D** easy to spot as the right choice. The other options will not make an oscillator.

G7B08 How is the efficiency of an RF power amplifier determined?
    A. Divide the DC input power by the DC output power
    B. Divide the RF output power by the DC input power
    C. Multiply the RF input power by the reciprocal of the RF output power
    D. Add the RF input power to the DC output power

We typically determine the efficiency by taking the ratio of the desired output to the supplied input. This computation means we can eliminate Answers C and D because they use incorrect math. Since this is RF efficiency, we look for the choice with RF power on the output side, which corresponds to **Answer B**. Answer A does not have the correct parameters for the computation.

G7B09 What determines the frequency of an LC oscillator?
   A. The number of stages in the counter
   B. The number of stages in the divider
   C. The inductance and capacitance in the tank circuit
   D. The time delay of the lag circuit

An *LC* oscillator is an analog circuit, not a digital circuit, so the counters and dividers of Answers A and B cannot be correct. The electrically-correct reasoning is that the values for the inductor and capacitor in the tank circuit set the oscillation frequency, so **Answer C** is the correct choice. Answer D is to distract you.

G7B10 Which of the following describes a linear amplifier?
   A. Any RF power amplifier used in conjunction with an amateur transceiver
   B. An amplifier in which the output preserves the input waveform
   C. A Class C high efficiency amplifier
   D. An amplifier used as a frequency multiplier

A linear amplifier preserves the input waveform, so **Answer B** is the right choice. Answer A is not a good choice because the question does not specify the type of amplifier. Answer C is a non-linear amplifier, and the question is asking about linear amplifiers. Answer D is technobabble.

G7B11 For which of the following modes is a Class C power stage appropriate for amplifying a modulated signal?
   A. SSB
   B. FM
   C. AM
   D. All of these choices are correct

The Class C amplifier can lead to a great deal of amplitude distortion in the output, so Answers A and C are not good choices because the modulated signal may become distorted. Also, Answer D cannot be correct. With proper construction, the transmitter can send a sinusoidal signal with minimal distortion, so it makes this amplifier a candidate for Frequency Modulation (FM). Remember, the FM waveform is a constant-amplitude sinusoid with the carrier restricted to vary over a relatively narrow range. **Answer B** is the best choice of those listed here.

7.5　G7C - RF Circuits

7.5.1　Overview

The *RF Circuits* question group in Subelement G7 quizzes you on RF-related components and techniques found in radio circuits. The *RF Circuits* group covers topics such as

- Receivers and transmitters
- Filters
- Oscillators

The test producer will select one of the 16 questions in this group for your exam.

7.5.2　Questions

G7C01 Which of the following is used to process signals from the balanced modulator then send them to the mixer in some single sideband phone transmitters?

    A. Carrier oscillator
    B. Filter
    C. IF amplifier
    D. RF amplifier

As Figure 7.5 shows, a typical SSB transmitter has the carrier oscillator of Answer A as one of the inputs to the balanced modulator along with the audio signal, so this is not a good choice because the question deals with the output of the balanced modulator. The RF amplifier of Answer D is one of the last components of the transmitter, and there needs to be some signal processing before this stage. This signal processing occurs in the middle of the transmitter. You need to remember that the correct name for this component is the filter, as in **Answer B**. Generally, the receiver has the IF amplifier, not the transmitter, so Answer C is not a good choice here.

G7C02 Which circuit is used to combine signals from the carrier oscillator and speech amplifier then send the result to the filter in some single sideband phone transmitters?

    A. Discriminator
    B. Detector
    C. IF amplifier
    D. Balanced modulator

Since we are still dealing with the SSB transmitter, we can eliminate the three components found in receivers as answers to this question. The discriminator, the detector, and the IF amplifier are all parts of receivers. You need to remember that the balanced modulator from **Answer D** is the appropriate component since it takes the audio signal and the carrier signal and combines them, as shown in Figure 7.5.

G7C03 What circuit is used to process signals from the RF amplifier and local oscillator then send the result to the IF filter in a superheterodyne receiver?
A. Balanced modulator
B. IF amplifier
C. Mixer
D. Detector

With a CW receiver, we can eliminate the transmitter component, which is the balanced modulator in Answer A. Referring to Figure 7.6, we see that the detector is the last stage in the receiver process, so Answer D is incorrect. The question is asking which component translates the signals from the RF stage down to the IF stage in the receiver. The mixer, as in **Answer C**, will have RF and local oscillator signals as its input and send its output to the IF filter. The IF amplifier, mentioned in Answer B, comes after this stage.

G7C04 What circuit is used to combine signals from the IF amplifier and BFO and send the result to the AF amplifier in some single sideband receivers?
A. RF oscillator
B. IF filter
C. Balanced modulator
D. Product detector

We are still looking at the operation of the receiver in Figure 7.6 here, so we can eliminate the balanced modulator of Answer C and the RF oscillator of Answer A. This process is near the last stage in the receiver process because we wish to amplify the Audio Frequency (AF) signals next. The product detector in **Answer D** is the correct choice since it has the IF and BFO signals for inputs and an audio-type signal for the output. Answer B comes before this stage, so it is incorrect as well.

G7C05 Which of the following is an advantage of a direct digital synthesizer (DDS)?
A. Wide tuning range and no need for band switching
B. Relatively high-power output
C. Relatively low power consumption
D. Variable frequency with the stability of a crystal oscillator

DDS is a new way of designing radio equipment where designers use digital circuits to process radio signals in place of using resistors, capacitors, and inductors to make the filter and oscillator circuits. This design means that the radio can modify the signals to have variable carrier frequencies that are also very stable. **Answer D** is the correct choice. Answers A, B, and C are incorrect statements.

G7C06 What should be the impedance of a low-pass filter as compared to the impedance of the transmission line into which it is inserted?
- A. Substantially higher
- B. About the same
- C. Substantially lower
- D. Twice the transmission line impedance

From our knowledge of maximum power transfer, we get the best results when we match the impedances, so **Answer B** is the best choice. All other options will not result in optimal power transfer, so they are not good choices.

G7C07 What is the simplest combination of stages that implement a super-heterodyne receiver?
- A. RF amplifier, detector, audio amplifier
- B. RF amplifier, mixer, IF discriminator
- C. HF oscillator, mixer, detector
- D. HF oscillator, prescaler, audio amplifier

To make a superheterodyne receiver, you must have an oscillator and a mixer to bring the RF signals down to the IF region, and then a detector to recover the signals. **Answer C** provides the minimum configuration. Answer A is missing the oscillator and the mixer stages. Answer B is missing the oscillator, and the discriminator means that it can only detect FM. Answer D is missing the mixer and the detector stages.

G7C08 What circuit is used in analog FM receivers to convert IF output signals to audio?
- A. Product detector
- B. Phase inverter
- C. Mixer
- D. Discriminator

Analog FM receivers use discriminators to convert from the IF to audio. This design makes **Answer D** the right choice. None of the components in Answers A, B, or C detect FM.

G7C09 What is the phase difference between the I and Q signals that software-defined radio (SDR) equipment uses for modulation and demodulation?
- A. Zero
- B. 90 degrees
- C. 180 degrees
- D. 45 degrees

Radio engineers call the I and Q channels "in quadrature," which means they are 90° apart. **Answer B** is the right choice, and the others are to distract you.

G7C10 What is an advantage of using I and Q signals in software-defined radios (SDRs)?
- A. The need for high resolution analog-to-digital converters is eliminated
- B. All types of modulation can be created with appropriate processing
- C. Minimum detectible signal level is reduced
- D. Converting the signal from digital to analog creates mixing products

The SDR is programmable, which allows the user to select different modulation formats by manipulating the I and Q signals with the software. This property makes **Answer B** the correct choice. The other options are incorrect statements.

G7C11 What is meant by the term "software defined radio" (SDR)?
- A. A radio in which most major signal processing functions are performed by software
- B. A radio that provides computer interface for automatic logging of band and frequency
- C. A radio that uses crystal filters designed using software
- D. A computer model that can simulate performance of a radio to aid in the design process

SDRs are another new technology revolutionizing radio design. **Answer A** shows how they got that name: the primary signal processing is done in software rather than with discrete components. Answers B and D are nice toys to have, but they have nothing to do with SDRs. Answer C is to distract you.

G7C12 What is the frequency above which a low-pass filter's output power is less than half the input power?
- A. Notch frequency
- B. Neper frequency
- C. Cutoff frequency
- D. Rolloff frequency

The cutoff frequency, as in **Answer C**, is the point where the output power is one-half the input power. This point is also called the $-3\,\mathrm{dB}$ point. The other choices are incorrect terminology.

G7C13 What term specifies a filter's maximum ability to reject signals outside its passband?
- A. Notch depth
- B. Rolloff
- C. Insertion loss
- D. Ultimate rejection

The ultimate rejection of **Answer D** specifies the maximum rejection ability. The notch depth determines the notch filter's attenuation, the rolloff is how quickly a filter transitions between passband and attenuation, and the insertion loss is the power loss from a signal going into a device.

G7C14 The bandwidth of a band-pass filter is measured between what two frequencies?
  A. Upper and lower half-power
  B. Cutoff and rolloff
  C. Pole and zero
  D. Image and harmonic

A Band Pass Filter (BPF) has both an upper and lower cutoff frequency (half-power frequencies), as in **Answer A**. Cutoff and rolloff are filter characteristics, but they are not correctly applied here. Pole and zero are terms used in describing the filter circuit design. Image and harmonic are signal processing terms for signals in the receiver.

G7C15 What term specifies a filter's attenuation inside its passband?
  A. Insertion loss
  B. Return loss
  C. Q
  D. Ultimate rejection

The attenuation in the passband is the insertion loss, which makes **Answer A** the correct choice. Based on the earlier question, we can eliminate ultimate rejection as a choice. The Q is the filter's quality factor, or how relatively narrow the passband is. The return loss is a signal power loss on a transmission line.

G7C16 Which of the following is a typical application for a Direct Digital Synthesizer?
  A. A high-stability variable frequency oscillator in a transceiver
  B. A digital voltmeter
  C. A digital mode interface between a computer and a transceiver
  D. A high-sensitivity radio direction finder

The Direct Digital Synthesizer can make a highly-stable, variable-frequency oscillator as in **Answer A**. Even though they have "digital" in the name, Answers B and C are not appropriate uses. The radio direction finder can be achieved with appropriate antennas and analog radio components too.

Chapter 8

G8 — SIGNALS AND EMISSIONS

8.1 Introduction

We saw the fundamentals of analog and digital modulation with the Technician Class license studies. To be proficient at operating on the amateur bands, we need to go deeper into both the analog and digital modes and their characteristics. The *Signals and Emissions* subelement for the General Class examination has the following question groups:

   A. Carriers and Modulation
   B. Frequency mixing
   C. Digital emission modes

Subelement 8 will generate three questions on the General Class examination.

8.2 Radio Engineering Concepts

Analog Modulation In the Technician Class study guide, we saw the three classes of analog modulation: Amplitude Modulation (AM), Frequency Modulation (FM), and Phase Modulation (PM). As a review, Table 8.1 lists the definitions and the envelope characteristics for each of these modulation types. Also, Figure 8.1 illustrates the resulting AM, FM, and PM carrier signals for the analog baseband input signal.

When the input signal is digital, and not a continuous analog signal, the amplitude, frequency, and phase states are limited to specific values. The modulation is called Amplitude Shift Keying (ASK), Frequency Shift Keying (FSK), and Phase Shift Keying (PSK), respectively, in this case.

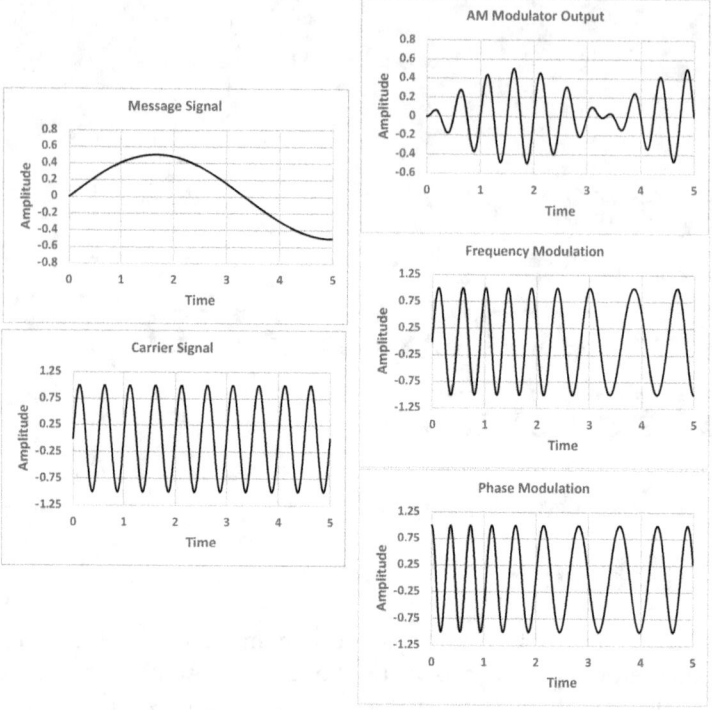

Figure 8.1: Carrier, baseband modulating signal, and the resulting AM, FM, and PM signals.

Table 8.1: Characteristics of analog modulation types.

Type	Definition	Envelope Characteristic
AM	input signal modifies the carrier's amplitude	time-varying
FM	input signal modifies the carrier's frequency	constant envelope
PM	input signal modifies the carrier's phase	constant envelope

Signal Bandwidth   As you remember from the Technician Class studies, the frequency space that we need to transmit a signal depends on the signal's underlying characteristics and the transmitter's modulation mode. To help with the questions on the General Class examination, we need to refer to Table 8.2 which lists typical analog bandwidth estimates for transmitting various signals both as generated at *baseband* (unmodulated) and as modulated signals. For digital bandwidth estimates, the bandwidth is typically 1 Hz per symbol per second of digital data transmission rate.

Table 8.2: Baseband and Modulated Bandwidth Estimates.

Baseband Signal	Baseband Bandwidth
CW	150 Hz
Voice	2 kHz to 3 kHz
Digital Signals	Approx. the transmission rate (symbols/second)
Slow-scan TV	3 kHz
Fast-Scan TV	6 MHz
Modulated Type	Modulated Signal Bandwidth
Single Sideband Voice	2 kHz to 3 kHz
Single Sideband Digital	Approx. the baseband bandwidth
Dual Sideband Voice	6 kHz
Narrowband FM Voice	15 kHz
Commercial Broadcast FM	75 kHz

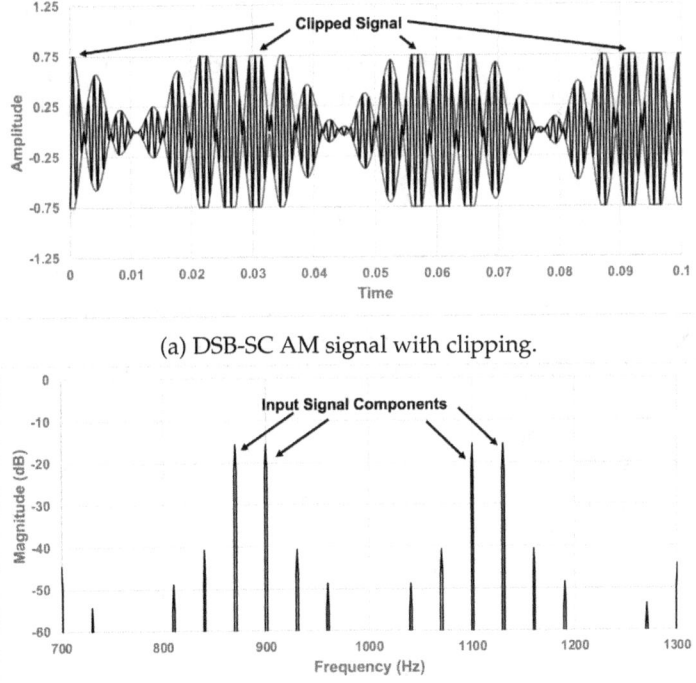

(a) DSB-SC AM signal with clipping.

(b) Resulting spectrum for the clipped AM signal.

Figure 8.2: Effects of clipping on an AM signal in the time and frequency domain.

Clipping and Flat-topping   The amplitude of the input signal and the linearity of any output amplifiers strongly influence the envelope power for AM carriers. Many transmitters clip the input signal to prevent excessive signal levels or *overmodulation*. The clipping comes with a price, and Figure 8.2 illustrates what happens. The input signal is two sinusoidal tones at 100 Hz and 130 Hz. The AM modulator has a carrier frequency of 1000 Hz. The transmitter clips the input signal at the 75 % level. Figure 8.2a shows the AM output envelope is cut to 75 % of its full-range value due to the input signal clipping process. This clipping produces spurious signal harmonics in the transmitted signal, as the bottom half of Figure 8.2b shows. This spectrum shows the input signal components at 870, 900, 1100, and 1130 Hz, while the other components come from the clipping process and are unwanted in the AM signal. These components produce excessive transmission bandwidth and distortion in the received AM signal.

A related issue is when the Radio Frequency (RF) amplifier has a large-amplitude signal on its input. The amplifier's range may be too small to accommodate the input signal, and the amplifier begins a non-linear distortion, limiting the output to a fixed, final value. This limiting flat-tops the signal, and the envelope will look like the illustration in Figure 8.2 as well.

Mixing and Image Frequencies   If you have had a class in trigonometry at some point, you probably needed to learn various trigonometric identities. One was for multiplying two sinusoids together. If ω_1 is the first sinusoid and ω_2 is the second, then

$$\cos \omega_1 \cos \omega_2 = {}^1/_2 \left(\cos \left(\omega_1 + \omega_2 \right) + \cos \left(\omega_1 - \omega_2 \right) \right)$$

This identity is a fundamental relationship in radio engineering! Figures 7.5 and 7.6 in Chapter 7 showed this in action. Here is how it works in a receiver: take an example carrier at 14.134 MHz and mix it with a Variable Frequency Oscillator (VFO) at 14.589 MHz. The mixing produces one component at 455 kHz and one component at 28.723 MHz. The 455 kHz forms the receiver's Intermediate Frequency (IF) component, and the other is filtered out.

Suppose another signal enters the mixer at 15.044 MHz. It will have a component also appearing at 455 kHz. This component is called the *image frequency* in the mixing process. An image frequency is any input frequency that, when mixed with the VFO frequency, will produce a component at the desired IF frequency.

Carson's Rule   Amplitude modulation is easy because the signal bandwidth on the channel is directly proportional to the message signal's bandwidth. FM and PM are not that simple because they represent non-linear processes. Engineers estimate the occupied bandwidth, B, of a FM transmission by Carson's rule in terms of the maximum *frequency deviation* (the maximum difference the modulator moves the carrier from its unmodulated frequency), D, and the

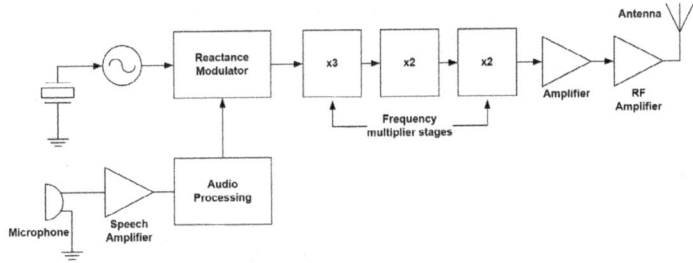

Figure 8.3: Block diagram for the elements of a narrow-band FM modulator.

bandwidth of the input message signal, W. We estimate the carrier's bandwidth as

$$B = 2(D + W)$$

If the input is a single tone, then W is the tone's frequency, f_m.

FM Modulators FM transmitters for voice signals, like those used in amateur radio communications, are often built around simple modulation systems like the one shown in Figure 8.3. Here a phone signal is used to modulate a simple reactance modulator. The output of that modulator is then multiplied through several stages to produce the signal at the desired frequency and with the proper FM characteristics. This output does not have the full quality of a commercial broadcast transmitter but transmits phone at an acceptable quality and price. You can even hold the transmitter in your hand.

Text Streaming and Data Packets Digital transmissions often need a bit more structure than analog transmissions like phone. The easiest digital transmission mode is text streaming. This mode is found in Continuous Wave (CW) transmission as well as in data transmissions, such as PSK31 or Radio TeleType (RTTY). The transmitter sends the digital characters as the user enters them in these modes. Any gaps or spaces flow as part of the transmission. Operators use these modes in point-to-point communications between two stations.

Other digital data modes use a formal packet structure as a part of a transmission protocol. This structure is needed when the data takes several hops between source and destination over a transmission network. The packet structure aids in routing the data over the network and provides additional information in the packet header to manage the transmission. The specific packet structure depends on the protocol definitions the designers construct. Usually, a protocol specification document details how the designers configured the software to build and use the packets. Figure 8.4 illustrates the components that might be in a data packet protocol. Usually, the data/text area is the majority of the packet.

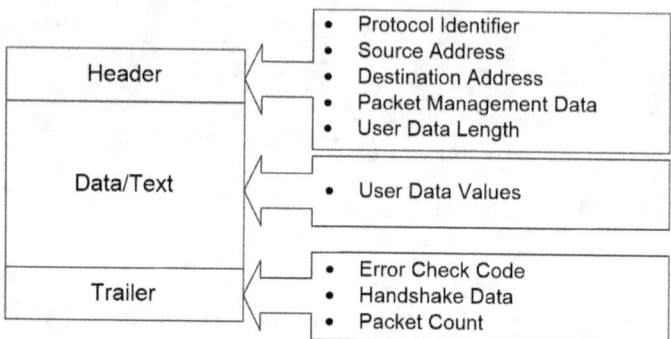

Figure 8.4: Components typically found in a data packet protocol. Specific details are given in the protocol documentation.

8.3   G8A - Carriers and Modulation

8.3.1   Overview

The *Carriers and Modulation* question group in Subelement G8 tests you on modulation-related definitions and adverse effects of the radio circuits on the modulation. The *Carriers and Modulation* group covers topics such as
- AM
- FM
- Single sideband
- Modulation envelope
- Digital modulation
- Overmodulation

The test producer will select one of the 12 questions in this group for your exam.

8.3.2   Questions

G8A01 How is an FSK signal generated?
- A. By keying an FM transmitter with a sub-audible tone
- B. By changing an oscillator's frequency directly with a digital control signal
- C. By using a transceiver's computer data interface protocol to change frequencies
- D. By reconfiguring the CW keying input to act as a tone generator

FSK means that the transmitter changes the carrier frequency based on the digital data value, as in **Answer B**, so that is the correct choice. Be careful with Answer C because the computer data interface is where the transmitted data enters the transmitter, but it does not describe the modulation process, so it is incorrect. Answers A is for accessing a repeater, and Answer D is technobabble.

G8A02 What is the name of the process that changes the phase angle of an RF signal to convey information?
A. Phase convolution
B. Phase modulation
C. Phase transformation
D. Phase inversion

Here we have two questions covering the names for modulation types. If you read the question carefully, you should be able to spot the answer quickly. We can write the carrier signal as $A \cos \Theta$ before modulation. If we change the phase angle of the carrier, Θ, in response to the message signal, then we must have phase modulation or **Answer B** — clever, those engineers! Again, watch out for something that appears similar. Answer A has "phase" in it, but there is no such thing as phase convolution modulation. Answers C and D are technobabble in this context.

G8A03 What is the name of the process that changes the instantaneous frequency of an RF wave to convey information?
A. Frequency convolution
B. Frequency transformation
C. Frequency conversion
D. Frequency modulation

This question is another one about names. Again, we can write the carrier signal as $A \cos \Theta$ before modulation. The time rate of change in the carrier phase, Θ, is called the carrier frequency. If we change the frequency in response to the message, we must have frequency modulation or **Answer D**. Again, watch out for the choices that look close but are wrong.

G8A04 What emission is produced by a reactance modulator connected to a transmitter RF amplifier stage?
A. Multiplex modulation
B. Phase modulation
C. Amplitude modulation
D. Pulse modulation

You need to remember that a reactance modulator will produce phase modulation, so the correct answer among these choices is **Answer B**. Multiplex modulation is used in television systems and not radio modulation, so it is incorrect. Amplitude modulation and pulse modulation come from different types of modulators, so those answers are also incorrect.

G8A05 What type of modulation varies the instantaneous power level of the RF signal?
   A. Frequency shift keying
   B. Phase modulation
   C. Frequency modulation
   D. Amplitude modulation

This question is about a different way of describing the AM in **Answer D**. Normally, we discuss how the input signal modifies the carrier signal, and this question states the result of that process. The transmitter modifies the RF envelope in response to the input signal. FM, PM, and FSK usually maintain constant RF envelope power, so these are incorrect choices.

G8A06 Which of the following is characteristic of QPSK31?
   A. It is sideband sensitive
   B. Its encoding provides error correction
   C. Its bandwidth is approximately the same as BPSK31
   D. All these choices are correct

QPSK31 is like sending two channels of BPSK31 in quadrature: one each on the I and Q channels, as in Figure 7.7. Each statement in Answers A, B, and C is correct, so **Answer D** is the right choice.

G8A07 Which of the following phone emissions uses the narrowest bandwidth?
   A. Single sideband
   B. Double sideband
   C. Phase modulation
   D. Frequency modulation

Here we need to remember the relative bandwidths of the modulation format. Single Sideband (SSB) takes about the same bandwidth as the voice signal itself, around 3 to 4 kHz. Dual Sideband (DSB) takes twice this amount. PM and FM usually take around five times the SSB amount. Therefore, the correct choice is **Answer A**.

G8A08 Which of the following is an effect of overmodulation?
   A. Insufficient audio
   B. Insufficient bandwidth
   C. Frequency drift
   D. Excessive bandwidth

Overmodulation will distort the transmitted signal, as Figure 8.2 shows. One of these distortions shows up as the signal occupying more bandwidth than it should, so **Answer D** is the correct choice. The alternatives do not arise from overmodulating the signal, so they are incorrect choices.

G8A09 What type of modulation is used by the FT8 digital mode?
- A. 8-tone frequency shift keying
- B. Vestigial sideband
- C. Amplitude compressed AM
- D. Direct sequence spread spectrum

The FT8 protocol uses an eight-tone frequency shift keying modulation format making **Answer A** the right choice. Operators often use Vestigial Side Band (VSB) in amateur TV transmission. Direct Sequence Spread Spectrum (DSSS) is typically used with Binary Phase Shift Keying (BPSK) or Quadrature Phase Shift Keying (QPSK) transmission. Answer C is to distract you.

G8A10 What is meant by the term "flat-topping," when referring to a single sideband phone transmission?
- A. Signal distortion caused by insufficient collector current
- B. The transmitter's automatic level control (ALC) is properly adjusted
- C. Signal distortion caused by excessive drive
- D. The transmitter's carrier is properly suppressed

When a signal is flat-topped, it is usually because of input limitations, such as the input range for an amplifier. As we saw in Figure 8.2, the flat-topping in the carrier cuts the carrier's envelope at some value, which distorts the signal. This result makes **Answer C** the best choice. Answer A is electrically untrue. Answer B is false because the input level not being set correctly produces overmodulation. Answer D is wrong because flat-topping is not a carrier suppression issue.

G8A11 What is the modulation envelope of an AM signal?
- A. The waveform created by connecting the peak values of the modulated signal
- B. The carrier frequency that contains the signal
- C. Spurious signals that envelop nearby frequencies
- D. The bandwidth of the modulated signal

As Figures 5.2 and 8.2a show, the modulation envelope is a trace of the peak values of the carrier signal, so **Answer A** is the right choice. The other options are to distract you.

G8A12 Which of the following narrow-band digital modes can receive signals with very low signal-to-noise ratios?
- A. MSK144
- B. FT8
- C. AMTOR
- D. MFSK32

The FT8 protocol designers were specifically aiming for very low Signal-to-

Noise Ratio (SNR) reception conditions, so choose **Answer B**. MSK144 is part of the WSJT-X software suite that was designed for meteor scatter applications on the Very High Frequency (VHF) bands. The designers of the other modes did not specifically have this goal.

8.4   G8B - Frequency Concepts

8.4.1   Overview

The *Frequency Concepts* question group in Subelement G8 quizzes you on the operations of mixers and FM modulators found in radio circuits and signal bandwidth. The *Frequency Concepts* group covers topics such as
- Frequency mixing
- Multiplication
- Bandwidths of various modes
- Deviation
- Duty cycle

The test producer will select one of the 12 questions in this group for your exam.

8.4.2   Questions

G8B01 Which mixer input is varied or tuned to convert signals of different frequencies to an intermediate frequency (IF)?
- A. Image frequency
- B. Local oscillator
- C. RF input
- D. Beat frequency oscillator

Designers use a mixer to shift a signal from one location in the frequency spectrum to another. This question refers to the receiver in Figure 7.6. The mixing process forms the sum and the difference of the input frequencies. This processing is what is happening when the receiver shifts the RF input signal to the 455-kHz IF stage. The correct choice is the Local Oscillator (LO), sometimes also called the VFO, in **Answer B**. The image frequency is a result of IF processing, so it is incorrect. The RF input is the input signal that the LO shifts. The Beat Frequency Oscillator (BFO) comes after the IF stage.

G8B02 If a receiver mixes a 13.800 MHz VFO with a 14.255 MHz received signal to produce a 455 kHz intermediate frequency (IF) signal, what type of interference will a 13.345 MHz signal produce in the receiver?
- A. Quadrature noise
- B. Image response
- C. Mixer interference
- D. Intermediate interference

If you subtract 13.345 MHz from 13.8 MHz, you obtain 455 kHz. The 13.345 MHz signal represents an image frequency because it can also appear at the IF, so the correct choice is **Answer B**. The other choices may sound reasonable, but they are just to distract you.

G8B03 What is another term for the mixing of two RF signals?
A. Heterodyning
B. Synthesizing
C. Cancellation
D. Phase inverting

Unless you are a communications engineer, this may be a difficult term for you to remember: heterodyning, as in **Answer A**. Synthesizing, in Answer B, builds a waveform, so this is incorrect. Engineers do not use cancellation in this context, so Answer C is incorrect. Phase inverting is to distract you.

G8B04 What is the stage in a VHF FM transmitter that generates a harmonic of a lower frequency signal to reach the desired operating frequency?
A. Mixer
B. Reactance modulator
C. Pre-emphasis network
D. Multiplier

Figure 8.3 shows how FM transmissions are frequently made. After the modulation step in the phase modulator, the signal undergoes several frequency multiplication stages. The phase modulator may be a reactance modulator, as in Answer B. The audio processing after the microphone may include a pre-emphasis network. A mixer, of Answer A, takes a carrier and multiplies it by a modulating signal Designers do not use that with FM. None of these steps generates the harmonics for which the question is asking. The frequency multiplying steps generates the harmonics, so **Answer D** is the correct choice.

G8B05 What is the approximate bandwidth of a PACTOR-III signal at maximum data rate?
A. 31.5 Hz
B. 500 Hz
C. 1800 Hz
D. 2300 Hz

This question is asking about the PACTOR packet protocol characteristics, and the correct answer is 2300 Hz, as in **Answer D**. The 31.5 Hz of Answer A is a reasonable estimate for the PSK31 format, but it does not apply here.

G8B06 What is the total bandwidth of an FM phone transmission having 5 kHz deviation and 3 kHz modulating frequency?
  A. 3 kHz
  B. 5 kHz
  C. 8 kHz
  D. 16 kHz

Here we apply Carson's rule for estimating FM bandwidth to find the answer. The bandwidth is computed using the formula $B = 2(5\,\text{kHz} + 3\,\text{kHz}) = 16\,\text{kHz}$, so **Answer D** is the right choice. Answer C is wrong because it is missing the factor of 2. Answers A and B are the input values, not the output bandwidth, so they are incorrect.

G8B07 What is the frequency deviation for a 12.21 MHz reactance modulated oscillator in a 5 kHz deviation, 146.52 MHz FM phone transmitter?
  A. 101.75 Hz
  B. 416.7 Hz
  C. 5 kHz
  D. 60 kHz

This question assumes a FM transmitter model with the frequency multiplier, as Figure 8.3 shows. To find the frequency deviation of the reactance modulator, we need to look at the frequency ratio between the local oscillator driving the reactance modulator and the final transmission frequency to find out the size of the multiplier in the FM transmitter. The multiplier "multiplies up" the oscillator frequency and the frequency deviation from the reactance modulator to the final RF transmission values. Using the values given in the question, we find that the multiplication factor is 146.52 MHz ÷ 12.21 MHz = 12. The question is asking us to figure out what the peak deviation of the reactance modulator was if the RF output has a peak frequency deviation of 5 kHz. To find the reactance modulator peak deviation, we need to divide the final 5 kHz by 12, which gives us 416.7 Hz, as in **Answer B**.

G8B08 Why is it important to know the duty cycle of the mode you are using when transmitting?
  A. To aid in tuning your transmitter
  B. Some modes have high duty cycles which could exceed the transmitter's average power rating.
  C. To allow time for the other station to break in during a transmission
  D. The attenuator will have to be adjusted accordingly

Many transmitters cannot operate continuously for very long before they get too hot, so the duty cycle (ratio of time transmitting a signal to the total time) is essential for thermal control of the transmitter. This limitation makes **Answer B** the right choice for this question. Answer A is generally not true. If the transmitter is getting too hot from generating the modulation, an attenuator after this stage will not help. Answer C is a silly distraction answer.

G8B09 Why is it good to match receiver bandwidth to the bandwidth of the operating mode?
  A. It is required by FCC rules
  B. It minimizes power consumption in the receiver
  C. It improves impedance matching of the antenna
  D. It results in the best signal-to-noise ratio

The noise in the receiver is proportional to the bandwidth of the receiver's filter. Therefore, we do not want the bandwidth to be any wider than it needs to be to capture the signal. If the filter bandwidth is narrower than the signal, then the signal will be distorted. If the user matches the filter's bandwidth to the signal, there will be no excess noise, and the signal will have its available power enter the detector. This matching will give us the best possible SNR, as in **Answer D**. The statements in Answers A, B, and C are false.

G8B10 What is the relationship between transmitted symbol rate and bandwidth?
  A. Symbol rate and bandwidth are not related
  B. Higher symbol rates require wider bandwidth
  C. Lower symbol rates require wider bandwidth
  D. Bandwidth is always half the symbol rate

For most digital data systems, the transmission bandwidth is approximately 1 Hz per symbol per second of the data transmission rate. Therefore, the transmission bandwidth scales directly with the data transmission rate. **Answer B** is the correct choice. The other answers are untrue statements.

G8B11 What combination of a mixer's Local Oscillator (LO) and RF input frequencies is found in the output?
  A. The ratio
  B. The average
  C. The sum and difference
  D. The arithmetic product

The mixer's output at this stage will produce the sum and difference frequencies as in **Answer C**. The other choices are all electrically incorrect statements.

G8B12 What process combines two signals in a non-linear circuit or connection to produce unwanted spurious outputs?
A. Intermodulation
B. Heterodyning
C. Detection
D. Rolloff

The key to the correct answer is the statement that the signals are unwanted. In this case, heterodyning is incorrect because it produces the desired signal. The question is discussing unwanted intermodulation interference that circuits make. The intermodulation in **Answer A** is the correct choice. Detection is what a receiver does, and rolloff is a filter property, so both of these choices do not apply here.

8.5 G8C - Digital Emission Modes

8.5.1 Overview

The *Digital Emission Modes* question group in Subelement G8 tests you on digital modulation topics. The test producer will select one of the 14 questions in this group for your exam.

8.5.2 Questions

G8C01 On what band do amateurs share channels with the unlicensed Wi-Fi service?
A. 432 MHz
B. 902 MHz
C. 2.4 GHz
D. 10.7 GHz

Amateurs share the 2.4 GHz band with unlicensed Wi-Fi emissions, which makes **Answer C** the correct choice. 432 MHz is an Amateur Service band and not for Wi-Fi. The Amateur Service and the Industrial, Scientific, and Medical (ISM) users share the 902 MHz band. While the 2.4 GHz band is also an ISM band, 902 MHz is not used for Wi-Fi. Space activity and radio astronomy users often are on the 10.7 GHz band.

G8C02 Which digital mode is used as a low-power beacon for assessing HF propagation?
A. WSPR
B. Olivia
C. PSK31
D. SSB-SC

The WSPR, pronounced "whisper," mode is a low-power beacon mode for propagation assessment, so **Answer A** is correct. Olivia and PSK31 are for regular amateur QSOs. The SSB-SC choice is to distract you. More information is available at https://physics.princeton.edu//pulsar/K1JT/wspr.html.

G8C03 What part of a packet radio frame contains the routing and handling information?
  A. Directory
  B. Preamble
  C. Header
  D. Footer

As Figure 8.4 shows, the header of a packet contains the routing and handling information, so **Answer C** is the right choice to answer this question. Answers A, B, and D are not part of a typical packet structure, so they are here to provide distraction choices for this question.

G8C04 Which of the following describes Baudot code?
  A. A 7-bit code with start, stop and parity bits
  B. A code using error detection and correction
  C. A 5-bit code with additional start and stop bits
  D. A code using SELCAL and LISTEN

You need to remember that Baudot codes are 5-bit codes, so **Answer C** is the right choice. Answer A is ASCII coding. Answers B and D are incorrect for Baudot codes.

G8C05 In the PACTOR protocol, what is meant by a NAK response to a transmitted packet?
  A. The receiver is requesting the packet be retransmitted
  B. The receiver is reporting the packet was received without error
  C. The receiver is busy decoding the packet
  D. The entire file has been received correctly

A "NAK" is a Negative Acknowledgement for a packet. In other words, the receiver did not correctly decode the packet, and the transmitter needs to retransmit it. This retransmission is the sense of **Answer A**, which is the right choice. The others are to distract you.

G8C06 What action results from a failure to exchange information due to excessive transmission attempts when using PACTOR or WINMOR?
  A. The checksum overflows
  B. The connection is dropped
  C. Packets will be routed incorrectly
  D. Encoding reverts to the default character set

A "failure to exchange" message means that packets suffered a transmission loss. If this is too great, the protocol will cease working, and the connection is dropped, as in **Answer B**. The other choices are to distract you in this case.

G8C07 How does the receiving station respond to an ARQ data mode packet containing errors?
   A. It terminates the contact
   B. It requests the packet be retransmitted
   C. It sends the packet back to the transmitting station
   D. It requests a change in transmitting protocol

Automatic Repeat reQuest (ARQ) is a protocol whereby a digital mode automatically requests the retransmission of a data packet that the receiver did not decode correctly. This request to retransmit is the meaning behind **Answer B**, so that is the right choice.

G8C08 Which of the following statements is true about PSK31?
   A. Upper case letters are sent with more power
   B. Upper case letters use longer Varicode bit sequences and thus slow down transmission
   C. Error correction is used to ensure accurate message reception
   D. Higher power is needed as compared to RTTY for similar error rates

Of the choices given, only the statement that sending uppercase takes longer is a true statement, so **Answer B** is the right choice.

G8C09 What does the number 31 represent in "PSK31"?
   A. The approximate transmitted symbol rate
   B. The version of the PSK protocol
   C. The year in which PSK31 was invented
   D. The number of characters that can be represented by PSK31

The "31" in PSK31 is approximately equal to the transmitted symbol rate, so **Answer A** is the correct choice. The other options are incorrect statements.

G8C10 How does forward error correction (FEC) allow the receiver to correct errors in received data packets?
   A. By controlling transmitter output power for optimum signal strength
   B. By using the Varicode character set
   C. By transmitting redundant information with the data
   D. By using a parity bit with each character

Forward Error Correction (FEC) codes add additional information to the transmitted data to permit the correction algorithm to work. This information is what **Answer C** is stating. A parity bit can detect errors but not correct them, so

Answer D is not a correct choice. Answer A would be beneficial, but it does not guard against transmission errors. The Varicode character set does not prevent errors either.

G8C11 How are the two separate frequencies of a Frequency Shift Keyed (FSK) signal identified?
  A. Dot and Dash
  B. On and Off
  C. High and Low
  D. Mark and Space

Engineers frequently call the one and zero data values for digital transmission "mark" and "space," respectively, so **Answer D** is the right choice. One can use the other options to represent the one and zero values, but not in this context, so they are incorrect.

G8C12 Which type of code is used for sending characters in a PSK31 signal?
  A. Varicode
  B. Viterbi
  C. Volumetric
  D. Binary

As we saw above, PSK31 uses a Varicode, as in **Answer A**. A Viterbi code is a typical forward error correction code, while a binary code is a base-2 encoding of numbers into zeroes and ones. They are not the character encoding, so they are incorrect. Volumetric usually refers to some form of image display, not character transmission.

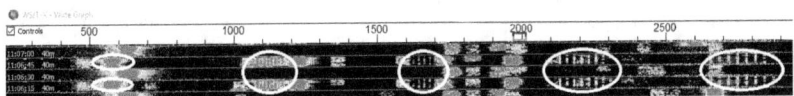

Figure 8.5: Overmodulation results for FT8 signals. The originating signals are on the left and the overmodulation signal products are the vertical lines to the right across the band.

G8C13 What is indicated on a waterfall display by one or more vertical lines on either side of a digital signal?
  A. Long path propagation
  B. Backscatter propagation
  C. Insufficient modulation
  D. Overmodulation

We saw waterfall displays in Chapter 2 with Figures 2.2 and 2.3. Figure 8.5

shows a waterfall with an example of FT8 overmodulation where the originating signals, just above 500 Hz, produce interfering signals across the band as marked by the ovals. Additionally, the originating FT8 signals are spreading out beyond their usual 50 Hz bandwidth. These examples of overmodulation are due to improper transmitter settings, so **Answer D** is the correct choice. The propagation path does not cause this, so Answers A and B are incorrect. Insufficient modulation is to distract you.

G8C14 Which of the following describes a waterfall display?
   A. Frequency is horizontal, signal strength is vertical, time is intensity
   B. Frequency is vertical, signal strength is intensity, time is horizontal
   C. Frequency is horizontal, signal strength is intensity, time is vertical
   D. Frequency is vertical, signal strength is horizontal, time is intensity

Here again, we refer to the waterfall displays in Chapter 2's Figures 2.2 and 2.3 and in Figure 8.5. From that, we can see that the waterfall has frequency along the horizontal axis, time in the vertical direction, and the intensity proportional to the signal strength, as in **Answer C**. The other choices are incorrect variations on this description, so be careful in reading the answer choices during the exam.

Chapter 9

G9 — ANTENNAS AND FEED LINES

9.1 Introduction

Because the privileges for the General Class license give operators more access to the High Frequency (HF) bands than they have with the Technician class license, many amateurs do not start using antennas and rigs for HF operation until they pass this examination. Your shack needs a good antenna system to support your operating modes. This subelement will test you on various antenna and feed line concepts, including the design equations and characteristics for several antenna types. This *Antennas and Feed Lines* subelement has the following question groups:

A. Antenna Feed Lines
B. Basic Antennas
C. Directional Antennas
D. Specialized Antennas

Subelement 8 will generate four questions on the General Class examination.

9.2 Radio Engineering Concepts

Transmission Lines Antennas and transceivers need transmission lines to carry the Radio Frequency (RF) signals flowing between them for transmission and reception. There are three general classes of transmission lines that you need to know about for the General Class license examination: parallel conductor line, air-insulated coaxial line, and dielectric-insulated coaxial cables. The fabricator produces *parallel conductor line* from two parallel wires with an air gap between them. The wire diameter, d, and the spacing, s, between the wires determine the impedance, Z, of the parallel conductor transmission line. You can compute

Table 9.1: Typical properties of twin-lead and co-axial cable including loss at 100 MHz

Type	Impedance Ω	Loss (dB /100 ft)
Twin lead	300	1.4
RG-6	75	1.8
RG-8	50	1.5
RG-8X	50	2.8
RG-11	75	1.6
RG-58	50	3.2
RG-59	75	2.5

the impedance using

$$Z = 276 \log \frac{2s}{d}$$

Coaxial cable has a center conductor surrounded by a braided ground wire with an insulator between the two. For *air-insulated co-axial cable*, the insulator is air. For *dielectric-insulated coaxial cable*, the insulator is usually a plastic material. We can compute the impedance from the inside diameter of the outer conductor, b, and the outside diameter of the inner conductor, a, using

$$Z = 138 \log \frac{b}{a}$$

Table 9.1 lists the properties for typical twin-lead and dielectric-insulated co-axial cables.

Impedance Matching   As we have seen several times in our studies, impedance matching is important in RF circuits because that is how we have maximum power transfer between devices such as transceivers and antennas. This concept appears in the General Class license exam questions as well. An important part of impedance matching is the Standing Wave Ratio (SWR) measured on the line. The SWR measures the transmitted and reflected power across the line. If the impedances match, the SWR is 1:1. If the impedances do not match, the ratio is >1:1. While one can compute the SWR from measured loss or the specific, measured electrical properties of the transmission lines, the SWR questions on the General Class examination reduce to

$$SWR = \frac{Z_1}{Z_2} : 1$$

where Z_1 is the larger impedance and Z_2 is the smaller impedance.

Antenna Gain and Patterns   Engineers use two common methods for describing the antenna: gain and radiation pattern. The antenna's *gain*, G, is a measure

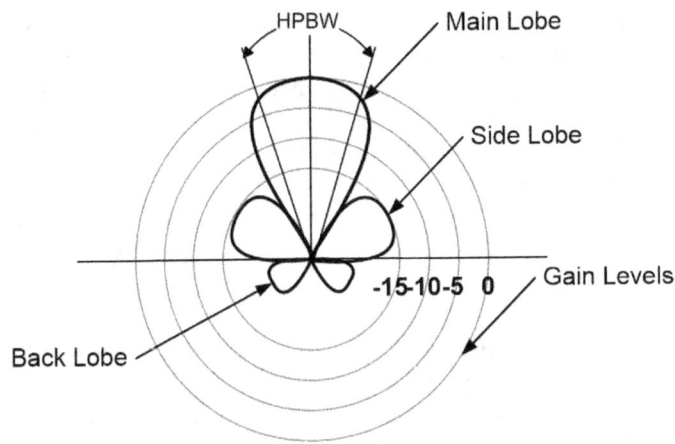

Figure 9.1: Representative antenna pattern plot.

of the strength of the emitted signal in a specified direction relative to a standard antenna. Typically, engineers use either an isotropic or dipole antenna as the reference standard. The *isotropic antenna* emits the RF radiation uniformly in all directions. We will describe the dipole antenna in a later paragraph. In either case, engineers usually express the gain measurement in decibel units. When referenced to an isotropic antenna, it is written as "dBi," while it is written as "dBd" when referenced to a dipole antenna.

The antenna pattern is a plot of the antenna's radiation emission, as Figure 9.1 illustrates. This pattern shows a two-dimensional slice of a three-dimensional figure. If you rotate the figure around the y-axis, you will see the full pattern. We assume the pattern works the same in the transmission and the reception directions. The pattern plot has several points of interest

Main Lobe — the part of the pattern where the highest intensity of the RF radiation appears; it points in the intended transmission direction when in use

Side Lobe — one of the secondary directions where the radiation still is transmitted or received

Back Lobe — radiation emitted or received in the opposite direction from the Main Lobe

Gain Levels — the strength of the radiation relative to the reference antenna in the same direction

Half Power Beamwidth — the angle enclosing the antenna pattern where the gain is within 3 dB of the maximum gain

Bandwidth — the frequency range where the antenna has a low SWR, typically less than 2:1

In addition to the Half Power Beam Width (HPBW), there are two important characteristics of the antenna pattern: the *front-to-back* gain ratio, and the *front-to-side* gain ratio. These measurements indicate the antenna's directionality. The

HPBW measures the main lobe's concentration in angular degrees. The two gain ratios measure how much radiation is emitted (or received) outside the main lobe through the its rear or sides. This measurement is usually in dBs.

There is an inverse relationship between gain and how sharply directional the pattern is. Low-gain antennas will behave more like isotropic antennas and have broad main lobes. High-gain antennas will have narrow main lobes. A high-gain, narrow angular width main lobe is a characteristic of a *directional antenna*. Because of antenna physics, one cannot have both high gain and a broad angular pattern.

Dipole Antennas   The dipole antenna is a simple conductor wire that is cut to match the radiation frequency, f, or wavelength, λ. Remember: $c = \lambda f$. Figure 9.2a illustrates that configuration. The conductor's length in feet, L_C, is normally $\lambda/2$, and is estimated from

$$L_C = \frac{468}{f(MHz)}$$

The dipole antenna's pattern looks like a figure-8 when observing it looking down the wire from the end. The gain is about 2.15 dB at maximum relative to the isotropic antenna.

There are several ways to mount dipole antennas, as Figure 9.3 illustrates. There are
  a. horizontal mounting with end and center supports
  b. an "Inverted V" with a center support and the antenna wire sloping down at an angle
  c. a "Sloper" with the whole antenna mounted at an angle from the vertical
  d. a "Lazy H" composed of two dipoles separated by ½-wavelength
The exact performance of the dipole will vary with configuration, near-by objects, and height above the ground.

Yagi Antennas   The Yagi (or, more formally, the Yagi-Uda) antenna builds on the dipole concept, and adds non-driven or *parasitic* elements to provide gain and directionality to the antenna. Figure 9.2b illustrates the 3-element Yagi antenna that is commonly used in amateur radio. For amateur use, the elements are normally mounted parallel to the Earth's surface. The elements are not simple wire, but usually are narrow-diameter, conductive metallic tubes placed along a non-conducting boom. Only one of the elements is fed by the transmission line, and it is basically a $\lambda/2$ dipole. The length of the fed element, L_F, the reflector element, L_R, and the director element, L_D, are approximated by

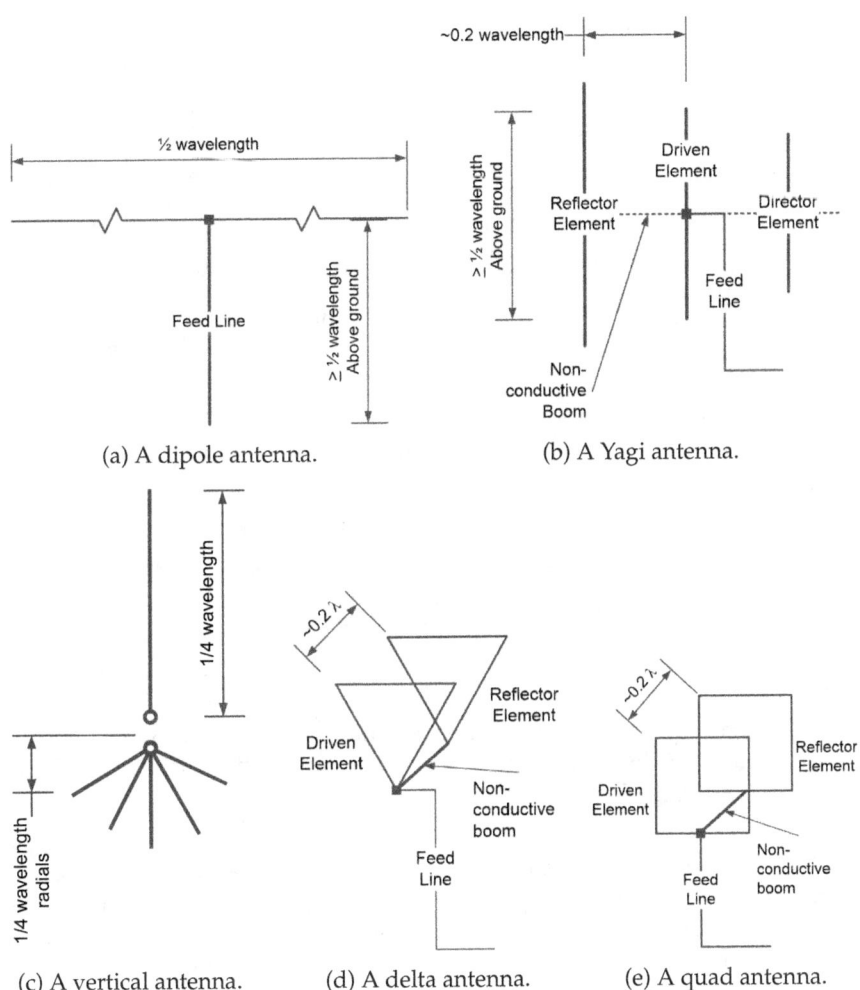

(a) A dipole antenna.   (b) A Yagi antenna.

(c) A vertical antenna.   (d) A delta antenna.   (e) A quad antenna.

Figure 9.2: Diagrams of representative antenna types.

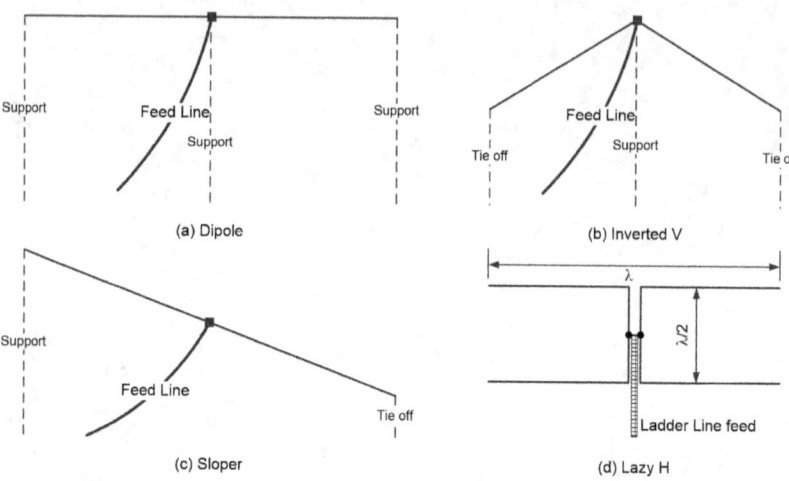

Figure 9.3: Common dipole mounting techniques.

$$L_F = \frac{468}{f(MHz)}$$

$$L_R = \frac{1.05 \times 468}{f(MHz)}$$

$$L_D = \frac{0.95 \times 468}{f(MHz)}$$

The 3-element Yagi antenna has a pattern similar to Figure 9.1. It has a maximum gain of around 7.5 dB relative to an isotropic antenna with a front-to-back ratio of approximately 25 dB.

Vertical Antennas   Fabricators construct vertical antennas using a vertical conductor over a horizontal ground plane, as Figure 9.2c shows. The typical vertical conductor is $\lambda/4$ in length. To have a good ground plane, the user typically installs a series of radial elements on or just under the top of the soil. The design equation for the conductor's length in feet is

$$L_C = \frac{234}{f(MHz)}$$

A typical $\lambda/4$ vertical has a gain of 3 dB.

Delta and Quad Antennas   The delta and quad loop antennas are common designs that use the Yagi antenna's driven element/reflector element method.

Figure 9.2 illustrates both antennas. In both cases, the total length, L_F, of the fed element is approximately one wavelength long with a design equation of

$$L_F = \frac{1005}{f(MHz)}$$

The reflected element is placed in-line with the driven element, and approximately 0.2λ apart. The total reflector element length, L_R, is longer than the driven element and is given by

$$L_R = \frac{1030}{f(MHz)}$$

The side lengths are proportional fractions (1/3 for delta and ¼ for quad) of the full length.

9.3   G9A - Antenna Feed Lines

9.3.1   Overview

The *Antenna Feed Lines* question group in Subelement G9 quizzes you on the properties of transmission lines and impedance matching for antennas. The *Antenna Feed Lines* group covers topics such as
- Antenna feed lines
- Characteristic impedance and attenuation
- SWR calculation, measurement, and effects
- Matching networks

The test producer will select one of the 13 questions in this group for your exam.

9.3.2   Questions

G9A01 Which of the following factors determine the characteristic impedance of a parallel conductor antenna feed line?
  A. The distance between the centers of the conductors and the radius of the conductors
  B. The distance between the centers of the conductors and the length of the line
  C. The radius of the conductors and the frequency of the signal
  D. The frequency of the signal and the length of the line

The impedance is a function of the geometry of the conductors, especially the spacing, not the length of the conductor or the operating frequency. We can eliminate Answers B, C, and D from further consideration from these properties. The distance between the conductors and their radius, as listed in **Answer A**, is the right choice.

G9A02 What are the typical characteristic impedances of coaxial cables used for antenna feed lines at amateur stations?

A. 25 and 30 ohms
B. 50 and 75 ohms
C. 80 and 100 ohms
D. 500 and 750 ohms

As we saw in Table 9.1, the impedances for coaxial cable typically are 50 Ω and 75 Ω, so **Answer B** is the correct choice. The other answers are not typical impedances for coaxial cable used in amateur rigs. Read the answers carefully because Answer D has numbers like the correct ones, but off by a factor of 10.

G9A03 What is the typical characteristic impedance of "window line" parallel transmission line?

A. 50 ohms
B. 75 ohms
C. 100 ohms
D. 450 ohms

You need to remember that the window-line parallel transmission line has an impedance of 450 Ω, so **Answer D** is the correct choice. Answers A and B are common for coaxial cable, so they look familiar. Read the question carefully to ensure you choose the window line value, not the coaxial value. Answer C is to distract you.

G9A04 What might cause reflected power at the point where a feed line connects to an antenna?

A. Operating an antenna at its resonant frequency
B. Using more transmitter power than the antenna can handle
C. A difference between feed-line impedance and antenna feed-point impedance
D. Feeding the antenna with unbalanced feed line

This question poses another way of asking the impedance matching question. We should be able to spot the correct choice as the differing impedances in **Answer C**. Answer A is where you should be operating, so it is the wrong choice here. Answers B and D do not cause reflected power intrinsically, so they are not good choices either.

G9A05 How does the attenuation of coaxial cable change as the frequency of the signal it is carrying increases?

A. Attenuation is independent of frequency
B. Attenuation increases
C. Attenuation decreases
D. Attenuation reaches a maximum at approximately 18 MHz

The attenuation increases with frequency, so **Answer B** is the right choice. The other options are all untrue statements.

G9A06 In what units is RF feed line loss usually expressed?
A. Ohms per 1000 feet
B. Decibels per 1000 feet
C. Ohms per 100 feet
D. Decibels per 100 feet

We measure signal loss in decibels (dB), not ohms. If you purchase coaxial cable, you will see that the manufacturer typically rates it in dB of loss per 100 feet, so **Answer D** is the best choice. Typical antenna cable lengths to outdoor antennas are on the order of 100 feet, so this unit should be easy to remember. The other answers are not correct measures of the loss as used to rate coaxial cable under normal circumstances.

G9A07 What must be done to prevent standing waves on an antenna feed line?
A. The antenna feed point must be at DC ground potential
B. The feed line must be cut to a length equal to an odd number of electrical quarter wavelengths
C. The feed line must be cut to a length equal to an even number of physical half wavelengths
D. The antenna feed point impedance must be matched to the characteristic impedance of the feed line

The statements in Answers A, B, and C will do nothing to prevent standing waves. We know that impedance matching does prevent standing waves, so **Answer D** is the correct choice.

G9A08 If the SWR on an antenna feed line is 5 to 1, and a matching network at the transmitter end of the feed line is adjusted to 1 to 1 SWR, what is the resulting SWR on the feed line?
A. 1 to 1
B. 5 to 1
C. Between 1 to 1 and 5 to 1 depending on the characteristic impedance of the line
D. Between 1 to 1 and 5 to 1 depending on the reflected power at the transmitter

If the feed line is 1:1 at the one end and 5:1 at the other, then the resulting SWR on the feed line will be 5:1, as in **Answer B**.

G9A09 What standing wave ratio will result when connecting a 50 ohm feed line to a non-reactive load having 200 ohm impedance?
   A. 4:1
   B. 1:4
   C. 2:1
   D. 1:2

The ratio of the impedances is $200\,\Omega \div 50\,\Omega$ or 4:1, so the correct choice is **Answer A**. Answer B has the correct numbers but in the wrong order, so that option is incorrect. The other answers are incorrect because they represent math errors.

G9A10 What standing wave ratio will result when connecting a 50 ohm feed line to a non-reactive load having 10 ohm impedance?
   A. 2:1
   B. 50:1
   C. 1:5
   D. 5:1

The ratio of the impedances is $50\,\Omega \div 10\,\Omega$ or 5:1, so the correct choice is **Answer D**. Answer C has the correct numbers but in the wrong order, so that option is incorrect. The other answers are math errors.

G9A11 What standing wave ratio will result when connecting a 50 ohm feed line to a non-reactive load having 50 ohm impedance?
   A. 2:1
   B. 1:1
   C. 50:50
   D. 0:0

Here, the impedances are the same, so the SWR is 1:1, and the correct choice is **Answer B**. Be careful of Answer C because it uses the impedance values as the SWR ratio. The other answers are to distract you. Answer D is impossible.

G9A12 What is the interaction between high standing wave ratio (SWR) and transmission line loss?
   A. There is no interaction between transmission line loss and SWR
   B. If a transmission line is lossy, high SWR will increase the loss
   C. High SWR makes it difficult to measure transmission line loss
   D. High SWR reduces the relative effect of transmission line loss

If the transmission line is lossy, a high SWR increases the loss because not all of the transmitted power will reach the destination. This interaction makes **Answer B** correct. The other choices are untrue statements.

G9A13 What is the effect of transmission line loss on SWR measured at the input to the line?
   A. The higher the transmission line loss, the more the SWR will read artificially low
   B. The higher the transmission line loss, the more the SWR will read artificially high
   C. The higher the transmission line loss, the more accurate the SWR measurement will be
   D. Transmission line loss does not affect the SWR measurement

If the line is lossy, then the SWR measurement will appear to be lower because the reflected power will dissipate through those losses, which makes **Answer A** the right choice. The other choices are untrue statements.

9.4 G9B - Basic Antennas

9.4.1 Overview

The *Basic Antennas* question group in Subelement G9 tests you on basic antenna knowledge across several varieties. The test producer will select one of the 12 questions in this group for your exam.

9.4.2 Questions

G9B01 What is one disadvantage of a directly fed random-wire HF antenna?
   A. It must be longer than 1 wavelength
   B. You may experience RF burns when touching metal objects in your station
   C. It produces only vertically polarized radiation
   D. It is more effective on the lower HF bands than on the higher bands

Not all operators would consider the statements in Answers A, C, and D to be disadvantages. The RF burns mentioned in **Answer B** are a definite disadvantage for all operators, so it is the best choice to answer this question.

G9B02 Which of the following is a common way to adjust the feed-point impedance of a quarter wave ground-plane vertical antenna to be approximately 50 ohms?
   A. Slope the radials upward
   B. Slope the radials downward
   C. Lengthen the radials
   D. Shorten the radials

This question is another radio practice one. One makes the desired adjustment by sloping the radials down, so **Answer B** is the right choice.

G9B03 Which of the following best describes the radiation pattern of a quarter-wave, ground-plane vertical antenna?
  A. Bi-directional in azimuth
  B. Isotropic
  C. Hemispherical
  D. Omnidirectional in azimuth

You need to remember that this antenna will have a pattern that is omnidirectional in azimuth, so **Answer D** is the correct choice. The other answers are untrue statements for this type of antenna.

G9B04 What is the radiation pattern of a dipole antenna in free space in the plane of the conductor?
  A. It is a figure-eight at right angles to the antenna
  B. It is a figure-eight off both ends of the antenna
  C. It is a circle (equal radiation in all directions)
  D. It has a pair of lobes on one side of the antenna and a single lobe on the other side

Dipole antennas have figure-eight patterns, unlike the patterns in Answers C and D. The figure eights are at right angles to the antenna, so the correct choice is **Answer A**. Answer B is wrong because it has the figure-eight pattern coming off at the wrong angle, so read the options carefully.

G9B05 How does antenna height affect the horizontal (azimuthal) radiation pattern of a horizontal dipole HF antenna?
  A. If the antenna is too high, the pattern becomes unpredictable
  B. Antenna height has no effect on the pattern
  C. If the antenna is less than 1/2 wavelength high, the azimuthal pattern is almost omnidirectional
  D. If the antenna is less than 1/2 wavelength high, radiation off the ends of the wire is eliminated

The height does affect the pattern, so Answer B is incorrect. Answers A and D are not electrically valid statements, so they are wrong. The nearly-omnidirectional pattern of **Answer C** is the best choice for this question.

G9B06 Where should the radial wires of a ground-mounted vertical antenna system be placed?
  A. As high as possible above the ground
  B. Parallel to the antenna element
  C. On the surface of the Earth or buried a few inches below the ground
  D. At the center of the antenna

We often call these wires ground radials, so **Answer C** gives the correct statement about their placement. Answers A, B, and D are incorrect placement options.

G9B07 How does the feed-point impedance of a 1/2 wave dipole antenna change as the antenna is lowered below 1/4 wave above ground?
A. It steadily increases
B. It steadily decreases
C. It peaks at about 1/8 wavelength above ground
D. It is unaffected by the height above ground

The impedance changes, so Answer D is incorrect. You need to remember that the impedance decreases in this case, as given in **Answer B**. The other choices are wrong.

G9B08 How does the feed point impedance of a 1/2 wave dipole change as the feed point is moved from the center toward the ends?
A. It steadily increases
B. It steadily decreases
C. It peaks at about 1/8 wavelength from the end
D. It is unaffected by the location of the feed point

You need to remember that the impedance increases in this case, so **Answer A** is correct. The other choices are electrically untrue statements. The Off Center Fed Dipole (OCFD) design places the feedpoint approximately $\lambda/6$ from the $\lambda/2$ dipole's end and uses a balun to match the impedance with the feed line.

G9B09 Which of the following is an advantage of a horizontally polarized as compared to a vertically polarized HF antenna?
A. Lower ground reflection losses
B. Lower feed point impedance
C. Shorter Radials
D. Lower radiation resistance

Answers B, C, and D are not necessarily disadvantages without knowing more specifics about the station. However, lower ground reflection losses are something everyone can use, and **Answer A** is the best choice to answer the question.

G9B10 What is the approximate length for a 1/2 wave dipole antenna cut for 14.250 MHz?
A. 8 feet
B. 16 feet
C. 24 feet
D. 33 feet

Using $L_C = 468/f(MHz) = 468/14.25 = 32.8$ ft as the design equation for the length of a ½-wave dipole makes **Answer D** the correct computation. Answer B is for a ¼-wavelength dipole at this frequency.

G9B11 What is the approximate length for a 1/2 wave dipole antenna cut for 3.550 MHz?
   A. 42 feet
   B. 84 feet
   C. 132 feet
   D. 263 feet

Here we use the same design equation as the previous question: $L_C = 468/3.55 = 131.8$ ft. This computation matches **Answer C**.

G9B12 What is the approximate length for a 1/4 wave vertical antenna cut for 28.5 MHz?
   A. 8 feet
   B. 11 feet
   C. 16 feet
   D. 21 feet

Here we modify the design equation by cutting the 468 factor to 234 to account for going from ½ to ¼ wavelength. The computation is $L_C = 234/28.5 = 8.2$ ft. This result matches **Answer A**. Be careful, Answer C matches the ½-wave case.

9.5   G9C - Directional Antennas

9.5.1   Overview

The *Directional Antennas* question group in Subelement G9 quizzes you on various directional antenna topics. The test producer will select one of the 16 questions in this group for your exam.

9.5.2   Questions

G9C01 Which of the following would increase the bandwidth of a Yagi antenna?
   A. Larger-diameter elements
   B. Closer element spacing
   C. Loading coils in series with the element
   D. Tapered-diameter elements

Of the choices given, the element diameter controls the bandwidth, so the correct choice is **Answer A**. Generally, designers use traps, or loading coils, to give the antenna access to the bands, not set the bandwidth. Answer C may sound like something you have heard, but it is not the best choice. Answers B and D will not have the desired effect of changing the antenna's bandwidth.

G9C02 What is the approximate length of the driven element of a Yagi antenna?
A. 1/4 wavelength
B. 1/2 wavelength
C. 3/4 wavelength
D. 1 wavelength

If you look at the Yagi antenna's design equations, you will see that the driven segment of the antenna is approximately ½ wavelength long. This property makes **Answer B** the correct choice.

G9C03 How do the lengths of a three-element Yagi reflector and director compare to that of the driven element?
A. The reflector is longer, and the director is shorter
B. The reflector is shorter, and the director is longer
C. They are all the same length
D. Relative length depends on the frequency of operation

If you look at Figure 9.2b and the Yagi antenna's design equations, you will see that **Answer A** is the right choice. Be careful with Answer B because the question reverses ordering. Answers C and D are incorrect statements.

G9C04 How does antenna gain stated in dBi compare to gain stated in dBd for the same antenna?
A. dBi gain figures are 2.15 dB lower then dBd gain figures
B. dBi gain figures are 2.15 dB higher than dBd gain figures
C. dBi gain figures are the same as the square root of dBd gain figures multiplied by 2.15
D. dBi gain figures are the reciprocal of dBd gain figures + 2.15 dB

The "dBi" indicates the gain relative to an isotropic antenna with a gain of 0 dB. The "dBd" means the gain relative to a dipole antenna which has a gain of 2.15 dB relative to an isotropic antenna. This relationship makes the dBi gain figure 2.15 dB higher than a dBd figure, as in **Answer B**. Be careful because Answer A has the order reversed. Answers C and D are technobabble.

G9C05 How does increasing boom length and adding directors affect a Yagi antenna?
A. Gain increases
B. Beamwidth increases
C. Front-to-back ratio decreases
D. Front-to-side ratio decreases

If done correctly, this modification will increase the gain of the antenna, so **Answer A** is the right choice. We saw above how to increase bandwidth, so you should spot Answer B as incorrect. Both Answers C and D are untrue statements because these quantities will increase with more antenna elements added.

G9C06 What configuration of the loops of a two-element quad antenna must be used for the antenna to operate as a beam antenna, assuming one of the elements is used as a reflector?
   A. The driven element must be fed with a balun transformer
   B. There must be an open circuit in the driven element at the point opposite the feed point
   C. The reflector element must be approximately 5 percent shorter than the driven element
   D. The reflector element must be approximately 5 percent longer than the driven element

This configuration is analogous to the Yagi antenna, where the reflector is the largest element. Here, the reflector is approximately 5 % longer, as in **Answer D**. Be careful with Answer C because it is shorter than what is needed. The other choices are to distract you.

G9C07 What does "front-to-back ratio" mean in reference to a Yagi antenna?
   A. The number of directors versus the number of reflectors
   B. The relative position of the driven element with respect to the reflectors and directors
   C. The power radiated in the major radiation lobe compared to that in the opposite direction
   D. The ratio of forward gain to dipole gain

The front-to-back ratio has to do with antenna radiation patterns for any directional antenna type. Answers A and B are incorrect because they deal with construction details, not radiation pattern shapes. Front and back are the radiation directions, and, as you may suspect, they are in opposite directions, so the correct choice is **Answer C**. Answer D corresponds to the relative gains of two antenna families, which is not what the question asks. Notice that the front-to-back ratio concept can be applied to many antenna designs.

G9C08 What is meant by the "main lobe" of a directive antenna?
   A. The magnitude of the maximum vertical angle of radiation
   B. The point of maximum current in a radiating antenna element
   C. The maximum voltage standing wave point on a radiating element
   D. The direction of maximum radiated field strength from the antenna

The main lobe for any antenna is the maximum field strength direction indication, so the correct choice is **Answer D**. Answers A, B, and C are measurements that one could make, but they do not tell you where the antenna is pointing the radiation, so they are incorrect.

G9C09 How does the gain of two three-element, horizontally polarized Yagi antennas spaced vertically 1/2 wavelength apart typically compare to the gain of a single three-element Yagi?
  A. Approximately 1.5 dB higher
  B. Approximately 3 dB higher
  C. Approximately 6 dB higher
  D. Approximately 9 dB higher

The two antennas will give you gain, but how much is the question. In this configuration, you will get about another 3 dB, so **Answer B** is the right choice.

G9C10 Which of the following can be adjusted to optimize forward gain, front-to-back ratio, or SWR bandwidth of a Yagi antenna?
  A. The physical length of the boom
  B. The number of elements on the boom
  C. The spacing of each element along the boom
  D. All of these choices are correct

Each of the items in Answers A, B, and C is an element in the design of the Yagi antenna, so the best choice is **Answer D**.

G9C11 Which HF antenna would be the best to use for minimizing interference?
  A. A quarter-wave vertical antenna
  B. An isotropic antenna
  C. A directional antenna
  D. An omnidirectional antenna

We wish to confine the transmission to the narrowest beam angle (smallest HPBW) possible to minimize interference. A directional antenna, as in **Answer C**, works in this way, so it is the right choice. The other antenna choices will have a wider RF radiation distribution, so they are incorrect choices.

G9C12 Which of the following is an advantage of using a gamma match with a Yagi antenna?
  A. It does not require that the driven element be insulated from the boom
  B. It does not require any inductors or capacitors
  C. It is useful for matching multiband antennas
  D. All of these choices are correct

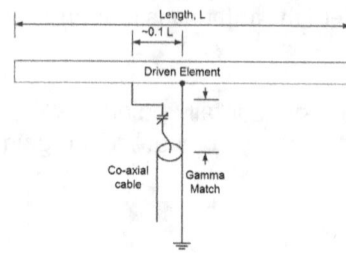

Figure 9.4: An antenna gamma match.

A gamma match may have a variable capacitor, so Answer B is incorrect. The matching network is generally for a single band, so Answer C is incorrect. Because of these statements, Answer D is also wrong. However, the fabricator does not need to insulate the driven elements from the boom, making **Answer A** the best choice.

G9C13 Approximately how long is each side of the driven element of a quad antenna?
  A. 1/4 wavelength
  B. 1/2 wavelength
  C. 3/4 wavelength
  D. 1 wavelength

As noted in our earlier discussion of delta and quad antennas, each side of the quad antenna shown in Figure 9.2e is approximately ¼ wavelength long. This length matches **Answer A**. Answer B is for the Yagi antenna. Answer D is the total length of the four sides of the quad antenna.

G9C14 How does the forward gain of a two-element quad antenna compare to the forward gain of a three-element Yagi antenna?
  A. About the same
  B. About 2/3 as much
  C. About 1.5 times as much
  D. About twice as much

This question asks about a "rule of thumb" relationship. You need to remember that gains are about the same, so **Answer A** is correct.

G9C15 What is meant by the terms dBi and dBd when referring to antenna gain?
  A. dBi refers to an isotropic antenna, dBd refers to a dipole antenna
  B. dBi refers to an ionospheric reflecting antenna, dBd refers to a dissipative antenna
  C. dBi refers to an inverted-vee antenna, dBd refers to a downward reflecting antenna
  D. dBi refers to an isometric antenna, dBd refers to a discone antenna

The term "dBi" is for decibels relative to an isotropic antenna, and "dBd" is for decibels relative to a dipole antenna, so **Answer A** is the right choice. The others are silly distraction answers.

G9C16 What is a beta or hairpin match?
  A. It is a shorted transmission line stub placed at the feed point of a Yagi antenna to provide impedance matching
  B. It is a ¼ wavelength section of 75 ohm coax in series with the feed point of a Yagi to provide impedance matching
  C. It is a series capacitor selected to cancel the inductive reactance of a folded dipole antenna
  D. It is a section of 300 ohm twinlead used to match a folded dipole antenna

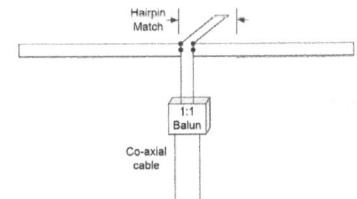

The short transmission line stub of **Answer A** is the description for a hairpin or beta match. The other choices are electrically incorrect statements for this type of matching circuit.

Figure 9.5: An antenna hairpin match.

9.6 G9D - Specialized Antennas

9.6.1 Overview

The *Specialized Antennas* question group in Subelement G9 tests you on specialized antenna topics. The test producer will select one of the 13 questions in this group for your exam.

9.6.2 Questions

G9D01 Which of the following antenna types will be most effective as a Near Vertical Incidence Skywave (NVIS) antenna for short-skip communications on 40 meters during the day?
  A. A horizontal dipole placed between 1/10 and 1/4 wavelength above the ground
  B. A vertical antenna placed between 1/4 and 1/2 wavelength above the ground
  C. A left-hand circularly polarized antenna
  D. A right-hand circularly polarized antenna

This antenna type is probably one to memorize: the low, horizontal dipole in **An-**

swer **A** is the correct description of the Near Vertical Incidence Skywave (NVIS) antenna. The other choices are to distract you when reading the question.

G9D02 What is the feed-point impedance of an end-fed half-wave antenna?
   A. Very low
   B. Approximately 50 ohms
   C. Approximately 300 ohms
   D. Very high

In the earlier question, we saw that the impedance increases as the antenna's feed moves from the center toward the end. The end-fed, half-wave antenna, also known as a "Zepp," has a very high input impedance, making **Answer D** the correct choice. The other options are to distract you.

G9D03 In which direction is the maximum radiation from a portable VHF/UHF "halo" antenna?
   A. Broadside to the plane of the halo
   B. Opposite the feed point
   C. Omnidirectional in the plane of the halo
   D. Toward the halo's supporting mast

The halo will produce an omnidirectional pattern in the halo's plane, so **Answer C** is correct. The other choices do not describe the halo's radiation pattern.

G9D04 What is the primary purpose of antenna traps?
   A. To permit multiband operation
   B. To notch spurious frequencies
   C. To provide balanced feed-point impedance
   D. To prevent out-of-band operation

Traps are inductors that allow the operator to tune the antenna across multiple bands. This purpose permits multiband operation, as in **Answer A**. The other choices do not describe the purpose of the trap.

G9D05 What is an advantage of vertical stacking of horizontally polarized Yagi antennas?
   A. It allows quick selection of vertical or horizontal polarization
   B. It allows simultaneous vertical and horizontal polarization
   C. It narrows the main lobe in azimuth
   D. It narrows the main lobe in elevation

Earlier, we saw that this configuration increases gain. In this case, stacking narrows the main lobe in elevation, so **Answer D** is the right choice.

G9D06 Which of the following is an advantage of a log periodic antenna?
A. Wide bandwidth
B. Higher gain per element than a Yagi antenna
C. Harmonic suppression
D. Polarization diversity

Log periodic antennas are known for having a wide bandwidth, so **Answer A** is the one to choose.

G9D07 Which of the following describes a log periodic antenna?
A. Element length and spacing vary logarithmically along the boom
B. Impedance varies periodically as a function of frequency
C. Gain varies logarithmically as a function of frequency
D. SWR varies periodically as a function of boom length

The antenna's name helps us make the right choice. The antenna's design places the elements at logarithmically-spaced intervals, and the element's length increases logarithmically, as in **Answer A**. The other choices are to distract you.

G9D08 How does a "screwdriver" mobile antenna adjust its feed-point impedance?
A. By varying its body capacitance
B. By varying the base loading inductance
C. By extending and retracting the whip
D. By deploying a capacitance hat

The operator changes the feed-point impedance by varying the screwdriver antenna's loading inductance, making **Answer B** the correct choice. Screwdriver antennas do not use the other options.

G9D09 What is the primary use of a Beverage antenna?
A. Directional receiving for low HF bands
B. Directional transmitting for low HF bands
C. Portable direction finding at higher HF frequencies
D. Portable direction finding at lower HF frequencies

Operators mostly use Beverage antennas for receiving on low HF bands, so **Answer A** is correct. Because they are very large antennas, they typically are not moved once installed, so they are not suitable for portable direction-finding.

G9D10 In which direction or directions does an electrically small loop (less than 1/3 wavelength in circumference) have nulls in its radiation pattern?
A. In the plane of the loop
B. Broadside to the loop
C. Broadside and in the plane of the loop
D. Electrically small loops are omnidirectional

An electrically small loop has nulls broadside to the loop, so **Answer B** is the correct choice. The other choices are electrically incorrect statements.

G9D11 Which of the following is a disadvantage of multiband antennas?
A. They present low impedance on all design frequencies
B. They must be used with an antenna tuner
C. They must be fed with open wire line
D. They have poor harmonic rejection

Because multiband antennas need to transmit across many amateur bands, and the amateur bands have harmonic relationships, designers cannot configure this antenna for harmonic suppression without compromising performance. This property makes **Answer D** the right choice. The other options are untrue statements.

G9D12 What is the common name of a dipole with a single central support?
A. Inverted V
B. Inverted L
C. Sloper
D. Lazy H

As Figure 9.3(a) shows, a typical dipole antenna requires three supports to keep it level and above the ground: one at each end and one in the center where the feed line attaches. In this question, the implication is that the two legs of the dipole are strung towards the ground rather than supported in line with the center point, as in Figure 9.3(b). In this case, they form a triangle with the center point at the apex, making an Inverted V configuration, as in **Answer A**. The Inverted L has the center support with one leg horizontal and held by a support, and one leg hanging down, so it is incorrect. The Sloper in Figure 9.3(c) has the legs of the dipole and the center in a straight line, but the whole antenna is rotated towards the ground at a 45° to 60° angle. A Lazy H in Figure 9.3(d) is an array of dipoles, and it is not a single dipole on one support.

G9D13 What is the combined vertical and horizontal polarization pattern of a multi-wavelength, horizontal loop antenna?
A. A figure-eight, similar to a dipole
B. Four major loops with deep nulls
C. Virtually omnidirectional with a lower peak vertical radiation angle than a dipole
D. Radiation maximum is straight up

This type of horizontal loop has an omnidirectional pattern with a lower peak vertical radiation angle as in **Answer C**. The other choices are incorrect descriptions.

Chapter 10

G0 — ELECTRICAL AND RF SAFETY

10.1 Introduction

Radio Frequency (RF) safety is a concern at all license levels. In the General Class license study, we go beyond the minimum safety rules and look into good engineering practices to keep you safe in the shack and to protect others around you. We will also further examine the effects of radiation on the human body. The *Electrical and RF Safety* subelement has the following question groups:

A. RF Safety Principles
B. Shack Safety

Subelement 10 will generate two questions on the General Class examination.

10.2 Radio Engineering Concepts

RF Safety Limits As we saw in the Technician Class study guide, industry and the government have established radiation limits for the RF exposure. Table 10.1 shows these limits. As we saw, the operator generally should perform a radiation analysis when a station is first established and then if they make any major equipment changes, such as a change in the antenna type or placement. The Federal Communications Commission (FCC) provides a 4-page worksheet at the end of the OET65 bulletin to assist with this evaluation. Many beginning operators will find that the worksheet shows their station does not need a full evaluation because they operate at a low power level. You can find more information on the Web at sites such as http://www.arrl.org/fcc-rf-expos ure-regulations-the-station-evaluation.

185

Table 10.1: Maximum Exposure Limits for Occupational/Controlled Exposure. (OET Bulletin 65, August 1997)

Frequency (MHz)	Electric Field Strength (E) (V/m)	Magnetic Field Strength (H) (A/m)	Power Density (S) (mW/cm^2)	Averaging Time (minutes)
0.3 – 3.0	614	1.63	100	6
3.0 – 30	$1842/f$	$4.89/f$	$900/f^2$	6
30 – 300	61.4	0.163	1.0	6
300 – 1500	—	—	$f/300$	6
1500 – 100,000	—	—	5	6

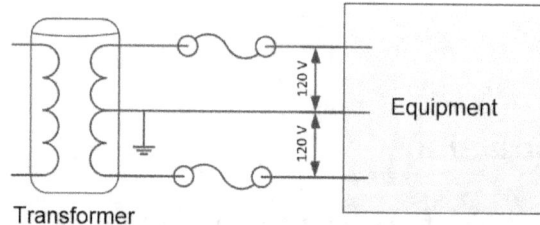

Figure 10.1: 240-V electrical wiring concept.

240-V Power   In the Technician Class study guide, we saw the standard Alternating Current (AC) wiring for 120 V. For high-power equipment, a 240 V connection is often used. Figure 10.1 illustrates the configuration. The 120-V AC uses one hot wire along with the neutral line. The 240-V AC uses both hot lines. Notice that the safe configuration uses fuses only on the two hot lines, not the neutral line.

Wire Gauges and Fuses   The current carrying capacity of a wire is an important safety consideration. If the wire does not have the current capacity rating for the circuit, it could overheat, leading to a safety hazard. Table 10.2 lists standard wire gauges and their rated current capacity when used as a single wire. Wire bundles need to use lower currents to keep from overheating.

Naturally, the circuit's fusing must also be compatible with the wire gauge. It is safe to have the fuse rated at a lower current capacity than the wire because the fuse will blow before the wire overheats. The reverse configuration is a problem because the wire could overheat before the fuse blows.

Table 10.2: Current carrying capacity of unbundled wires.

AWG Gauge	Rated Current (A)
22	3
20	5
16	15
14	15
12	20
10	30
8	40

10.3 G0A - RF Safety Principles

10.3.1 Overview

The *RF Safety Principles* question group in Subelement G0 concentrates on shack safety evaluation and exposure limits. The *RF Safety Principles* group covers topics such as
- RF Safety principles, rules, and guidelines
- Routine station evaluation

The test producer will select one of the 11 questions in this group for your exam.

10.3.2 Questions

G0A01 What is one way that RF energy can affect human body tissue?
A. It heats body tissue
B. It causes radiation poisoning
C. It causes the blood count to reach a dangerously low level
D. It cools body tissue

As you may suspect, Answer D is incorrect. Answer B is generally most concerned with effects caused by energetic particles and high-energy electromagnetic radiation rather than the lower-energy RF radiation found in amateur radio, so it is not the best choice. Answer C is also for ionizing radiation, not for the RF radiation used in amateur rigs. The best choice among the options given is the tissue heating in **Answer A**.

G0A02 Which of the following properties is important in estimating whether an RF signal exceeds the maximum permissible exposure (MPE)?
A. Its duty cycle
B. Its frequency
C. Its power density
D. All of these choices are correct

As we can see from the Maximum Permissible Exposure (MPE) information in Table 10.1, each of the factors listed in Answers A, B, and C is important, so the best choice is **Answer D**.

G0A03 [97.13(c)(1)] How can you determine that your station complies with FCC RF exposure regulations?
  A. By calculation based on FCC OET Bulletin 65
  B. By calculation based on computer modeling
  C. By measurement of field strength using calibrated equipment
  D. All of these choices are correct

The Part 97 rules state that before "causing or allowing an amateur station to transmit from any place where the operation of the station could cause human exposure to RF electromagnetic field levels in excess of those allowed under Sec. 1.1310 of this chapter, the licensee is required to take certain actions." The actions are to perform a station evaluation using one of the methods given in Answers A, B, and C, so **Answer D** is the best choice.

G0A04 What does "time averaging" mean in reference to RF radiation exposure?
  A. The average amount of power developed by the transmitter over a specific 24-hour period
  B. The average time it takes RF radiation to have any long-term effect on the body
  C. The total time of the exposure
  D. The total RF exposure averaged over a certain time

Answers A, B, C, and D may all look reasonable on the surface, but only **Answer D** contains the correct definition for time averaging for RF evaluation.

G0A05 What must you do if an evaluation of your station shows RF energy radiated from your station exceeds permissible limits?
  A. Take action to prevent human exposure to the excessive RF fields
  B. File an Environmental Impact Statement (EIS-97) with the FCC
  C. Secure written permission from your neighbors to operate above the controlled MPE limits
  D. All of these choices are correct

Answer B is incorrect because the FCC does not require an Environmental Impact Statement in this case. Answer C is incorrect because the government will not allow the neighbors to overrule the standard. Answer D is also wrong because Answers B and C are incorrect. **Answer A** provides the best response to this question because you must take action to correct the situation.

G0A06 What precaution should be taken when installing a ground-mounted antenna?
- A. It should not be installed higher than you can reach
- B. It should not be installed in a wet area
- C. It should limited to 10 feet in height
- D. It should be installed such that it is protected against unauthorized access

Since we are dealing with RF safety, only **Answer D** ensures that the system is safe. The other Answers do not address RF safety.

G0A07 What effect does transmitter duty cycle have when evaluating RF exposure?
- A. A lower transmitter duty cycle permits greater short-term exposure levels
- B. A higher transmitter duty cycle permits greater short-term exposure levels
- C. Low duty cycle transmitters are exempt from RF exposure evaluation requirements
- D. High duty cycle transmitters are exempt from RF exposure requirements

According to the exposure guidelines, answers C and D are untrue statements. Answer B is the opposite of the procedure, so it is out too. Because a lower duty cycle produces RF radiation for a shorter duration in each averaging period, **Answer A** is the correct choice for this question.

G0A08 Which of the following steps must an amateur operator take to ensure compliance with RF safety regulations when transmitter power exceeds levels specified in FCC Part 97.13?
- A. Post a copy of FCC Part 97.13 in the station
- B. Post a copy of OET Bulletin 65 in the station
- C. Perform a routine RF exposure evaluation
- D. Contact the FCC for a visit to conduct a station evaluation

Merely posting a copy of the rules does not ensure compliance, so Answers A and B are incorrect. Only the routine evaluation mentioned in **Answer C** might be a help, so it is the best choice for this question. The FCC does not do this evaluation — the station owner must perform the assessment.

G0A09 What type of instrument can be used to accurately measure an RF field?
- A. A receiver with an S meter
- B. A calibrated field strength meter with a calibrated antenna
- C. An SWR meter with a peak-reading function
- D. An oscilloscope with a high-stability crystal marker generator

An S meter is a crude relative measurement of the received signal strength, so Answer A is out. A Standing Wave Ratio (SWR) meter measures reflected power, not emitted power, so Answer C is out. Operators use the oscilloscope of Answer D to check the carrier's frequency, not the field strength, so it is incorrect. The correct way to make the measurement is with calibrated field strength meters and antenna, as given in **Answer B**.

G0A10 What is one thing that can be done if evaluation shows that a neighbor might receive more than the allowable limit of RF exposure from the main lobe of a directional antenna?
  A. Change to a non-polarized antenna with higher gain
  B. Post a warning sign that is clearly visible to the neighbor
  C. Use an antenna with a higher front-to-back ratio
  D. Take precautions to ensure that the antenna cannot be pointed in their direction

Changing the polarization will not change the power density radiated towards your neighbors, so Answer A is incorrect. A warning sign is insufficient, so Answer B is not a good choice. Answer C may increase the radiation towards the neighbors, so it is out too. **Answer D** is the correct choice for this question.

G0A11 What precaution should you take if you install an indoor transmitting antenna?
  A. Locate the antenna close to your operating position to minimize feed-line radiation
  B. Position the antenna along the edge of a wall to reduce parasitic radiation
  C. Make sure that MPE limits are not exceeded in occupied areas
  D. Make sure the antenna is properly shielded

Because this is an indoor antenna, we must be careful about exposure to the building occupants. We must ensure the station does not exceed the MPE limits, so **Answer C** is the best choice for this question. Answers A, B, and D are not concerned with protecting the occupants, so they are not good practice.

10.4   G0B - Shack Safety

10.4.1   Overview

The *Shack Safety* question group in Subelement G0 tests you on safe wiring practices for electrical circuits. The *Shack Safety* group covers topics such as
  • Station safety,
  • Electrical shock, safety grounding, and fusing
  • Interlocks
  • Wiring
  • Antenna and tower safety
The test producer will select one of the 14 questions in this group for your exam.

10.4.2 Questions

G0B01 Which wire or wires in a four-conductor connection should be attached to fuses or circuit breakers in a device operated from a 240 VAC single phase source?
A. Only the two wires carrying voltage
B. Only the neutral wire
C. Only the ground wire
D. All wires

The neutral and ground wires in Answers B and C should not be fused to ensure proper circuit operation when faults occur, so they are incorrect choices. Answer D is also wrong because it includes fusing the neutral and ground wires. The correct choice is the voltage-carrying wires of **Answer A**.

G0B02 According the National Electrical Code, what is the minimum wire size that may be used safely for wiring with a 20 ampere circuit breaker?
A. AWG number 20
B. AWG number 16
C. AWG number 12
D. AWG number 8

Based on Table 10.2, you need to remember that manufacturers rate the 12-gauge wire in **Answer C** for up to 20 A. Manufacturers rate the wire gauges listed in Answers A and B for lower currents, so they are not safe choices. The 8-gauge wire of Answer D will be able to conduct 20 A, but it can also handle much more than 20 A, so it is not the best choice in this context because it might permit you to carry much more current than is safe for your application.

G0B03 Which size of fuse or circuit breaker would be appropriate to use with a circuit that uses AWG number 14 wiring?
A. 100 amperes
B. 60 amperes
C. 30 amperes
D. 15 amperes

In Table 10.2, we see that manufacturers rate 14-gauge wire for 15 A at most, so **Answer D** is the right choice. Using any of those higher currents could be a safety hazard.

G0B04 Which of the following is a primary reason for not placing a gasoline-fueled generator inside an occupied area?
  A. Danger of carbon monoxide poisoning
  B. Danger of engine over torque
  C. Lack of oxygen for adequate combustion
  D. Lack of nitrogen for adequate combustion

A gasoline engine will emit carbon monoxide. If the generator is inside an enclosed area, the CO can build up to deadly levels. **Answer A** reflects this danger.

G0B05 Which of the following conditions will cause a Ground Fault Circuit Interrupter (GFCI) to disconnect the 120 or 240 Volt AC line power to a device?
  A. Current flowing from one or more of the voltage-carrying wires to the neutral wire
  B. Current flowing from one or more of the voltage-carrying wires directly to ground
  C. Overvoltage on the voltage-carrying wires
  D. All of these choices are correct

The Ground Fault Circuit Interrupter (GFCI) looks for current on the ground wire. This property makes **Answer B** the correct choice.

G0B06 Which of the following is covered by the National Electrical Code?
  A. Acceptable bandwidth limits
  B. Acceptable modulation limits
  C. Electrical safety inside the ham shack
  D. RF exposure limits of the human body

The National Electrical Code is concerned with electrical safety at all locations, not just the ham shack, so **Answer C** is still the right choice. The FCC is more concerned with the items in Answers A, B, and D.

G0B07 Which of these choices should be observed when climbing a tower using a safety belt or harness?
  A. Never lean back and rely on the belt alone to support your weight
  B. Confirm that the belt is rated for the weight of the climber and that it is within its allowable service life
  C. Ensure that all heavy tools are securely fastened to the belt D-ring
  D. All of these choices are correct

Answer B will keep the harness from failing in use, so this is the best answer here. Answer A is probably a good idea, but not as good as Answer B. Answer C is not a good choice because tools should be secured separately and not tied to your body. Answer D is not true because Answers A and C are not true.

G0B08 What should be done by any person preparing to climb a tower that supports electrically powered devices?
  A. Notify the electric company that a person will be working on the tower
  B. Make sure all circuits that supply power to the tower are locked out and tagged
  C. Unground the base of the tower
  D. All of these choices are correct

Answer B makes sure that all powered devices are disabled, so it is the best choice. Answers A and C will not prevent electrical hazards. Because they are not true, Answer D is also untrue.

G0B09 Which of the following is true of an emergency generator installation?
  A. The generator should be located in a well-ventilated area
  B. The generator must be insulated from ground
  C. Fuel should be stored near the generator for rapid refueling in case of an emergency
  D. All these choices are correct

Based on the earlier question, did you spot **Answer A** as the correct choice? Answers B and C could give rise to safety hazards, so they are incorrect, along with Answer D.

G0B10 Which of the following is a danger from lead-tin solder?
  A. Lead can contaminate food if hands are not washed carefully after handling the solder
  B. High voltages can cause lead-tin solder to disintegrate suddenly
  C. Tin in the solder can "cold flow," causing shorts in the circuit
  D. RF energy can convert the lead into a poisonous gas

Answers B, C, and D are all untrue. Lead contamination, as in **Answer A**, is correct. That is one reason the electronics industry is moving away from lead wherever possible.

G0B11 Which of the following is good practice for lightning protection grounds?
  A. They must be bonded to all buried water and gas lines
  B. Bends in ground wires must be made as close as possible to a right angle
  C. Lightning grounds must be connected to all ungrounded wiring
  D. They must be bonded together with all other grounds

Your station's design should bond the lightning protection grounds with other grounds, so **Answer D** is the right choice. Electrically, each of the other statements is untrue.

G0B12 What is the purpose of a power supply interlock?
   A. To prevent unauthorized changes to the circuit that would void the manu-facturer's warranty
   B. To shut down the unit if it becomes too hot
   C. To ensure that dangerous voltages are removed if the cabinet is opened
   D. To shut off the power supply if too much voltage is produced

The best reason is given in **Answer C** because that promotes electrical safety. Answer A is a good reason, but not as good as Answer C. Answer B is called a thermal fuse. Answer D is some form of an internal protection circuit.

G0B13 What must you do when powering your house from an emergency generator?
   A. Disconnect the incoming utility power feed
   B. Insure that the generator is not grounded
   C. Insure that all lightning grounds are disconnected
   D. All of these choices are correct

If you are trying to run your house from an emergency generator and the local power company at the same time, a dangerous interaction can occur. You can prevent this situation by disconnecting the utility power source, as in **Answer A**. Answers B and C are unsafe and will not prevent this problem. Since Answers B and C are incorrect, Answer D is also wrong.

G0B14 What precaution should you take whenever you adjust or repair an antenna?
   A. Ensure that you and the antenna structure are grounded
   B. Turn off the transmitter and disconnect the feed line
   C. Wear a radiation badge
   D. All of these choices are correct

For personal safety, the transmitter should be off and disconnected, making **Answer B** the correct choice. Neither Answer A nor Answer C will protect you from RF radiation, so they are not good choices. Answer D is also incorrect.

Appendix A

Acronyms, Abbreviations, and Symbols

A.1 Radio Acronyms and Abbreviations

AC   Alternating Current

AF   Audio Frequency

AFSK   Audio Frequency Shift Keying

ALC   Automatic Level Control

AM   Amplitude Modulation

AREDN   Amateur Radio Emergency Data Network

ARES   Amateur Radio Emergency Service

ARQ   Automatic Repeat reQuest

ASK   Amplitude Shift Keying

BFO   Beat Frequency Oscillator

BPF   Band Pass Filter

BPSK   Binary Phase Shift Keying

BJT   Bipolar Junction Transistor

CFR   Code of Federal Regulations

CMOS   Complementary Metal Oxide Semiconductor

CSCE   Certificate of Successful Completion of an Examination

CW	Continuous Wave
DC	Direct Current
DDS	Direct Digital Synthesis
DSB	Dual Sideband
DSB-RC	Dual Sideband - Residual Carrier
DSB-SC	Dual Sideband - Suppressed Carrier
DSP	Digital Signal Processor
DSSS	Direct Sequence Spread Spectrum
DX	Distant
FCC	Federal Communications Commission
FEC	Forward Error Correction
FET	Field Effect Transistor
FM	Frequency Modulation
FRN	FCC Registration Number
FSK	Frequency Shift Keying
GFCI	Ground Fault Circuit Interrupter
HF	High Frequency
HPBW	Half Power Beam Width
HPF	High Pass Filter
IARU	International Amateur Radio Union
IC	Integrated Circuit
IEEE	Institute of Electrical and Electronics Engineers
IF	Intermediate Frequency
ISM	Industrial, Scientific, and Medical
ITU	International Telecommunication Union
JFET	Junction Field Effect Transistor
LCD	Liquid Crystal Display
LED	Light Emitting Diode

LF	Low Frequency
LO	Local Oscillator
LPF	Low Pass Filter
LSB	Lower Side Band
LUF	Lowest Usable Frequency
MF	Medium Frequency
MFSK	Multiple Frequency Shift Keying
MMIC	Monolithic Microwave Integrated Circuit
MOSFET	Metal Oxide Semiconductor Field Effect Transistor
MPE	Maximum Permissible Exposure
MUF	Maximum Usable Frequency
NCVEC	National Conference of Volunteer Examiner Coordinators
NOAA	National Oceanic and Atmospheric Administration
NVIS	Near Vertical Incidence Skywave
OCFD	Off Center Fed Dipole
PEP	Peak Envelope Power
PEV	Peak Envelope Voltage
PM	Phase Modulation
PSK	Phase Shift Keying
PTT	Push to Talk
QPSK	Quadrature Phase Shift Keying
RACES	Radio Amateur Civil Emergency Service
RF	Radio Frequency
RIT	Receiver Incremental Tuning
RMS	Root Mean Square
ROM	Read Only Memory
RST	Readability-Signal Strength-Tone
RTTY	Radio TeleType

SDR	Software Defined Radio
SID	Sudden Ionospheric Disturbance
SNR	Signal-to-Noise Ratio
SS	Spread Spectrum
SSB	Single Sideband
SWR	Standing Wave Ratio
TTL	Transistor-Transistor Logic
UHF	Ultra High Frequency
ULS	Universal Licensing System
US	United States
USB	Upper Side Band
VE	Volunteer Examiner
VEC	Volunteer Examiner Coordinator
VFO	Variable Frequency Oscillator
VHF	Very High Frequency
VMP	Volunteer Monitoring Program
VOX	Voice Operated Switch
VSB	Vestigial Side Band
XCVR	Transceiver

A.2   Functions, Symbols, Units, and Variables

A	ampere
B	Transmitted Bandwidth, Hz
c	Speed of Light, $299\,792\,458$ m/s
C	Capacitance, F
D	Carrier Deviation, Hz
d	Wire diameter
dB	decibel

dB W	decibel Watt
dBm	decibel milliWatt
f	Frequency, Hz
f_m	Modulating Tone Frequency, Hz
F	farad
G	Gain, unitless
G	Conductance, S
GHz	1 000 000 000 Hz
H	henry
Hz	hertz
i	Current, A
kHz	1000 Hz
km	1000 m
λ	Wavelength, m
L	Inductance, H
LC	Inductor-Capacator
L_C	Conductor antenna length, ft
L_D	Director element length, ft
L_F	Feed element length, ft
L_R	Reflector element length, ft
m	meter
mA	0.001 A
μA	0.000 001 A
MHz	1 000 000 Hz
mH	0.001 H
mV	0.001 V
mW	0.001 W
μF	0.000 001 F

μH	0.000 001 H
μV	0.000 001 V
μW	0.000 001 W
Q	Quality Factor
P	Power, W
pF	0.000 000 000 001 F
R	Resistance, Ω
R_b	Bit Rate, bps
RC	Resistor-Capacitor
RLC	Resistor-Inductor-Capacitor
s	second
s	Wire spacing
S	Siemens
sps	Symbols per second
V	volt
W	Baseband Signal Bandwidth, Hz
Ω	ohm
W	watt
X	Reactance, Ω
Y	Admittance, S
Z	Impedance, Ω